Biochemical and Clinical Aspects of Pteridines
Volume 1
Cancer · Immunology · Metabolic Diseases

Biochemical and Clinical Aspects of Pteridines Volume 1

Cancer · Immunology Metabolic Diseases

Proceedings
First Winter Workshop on Pteridines
March 7-10, 1982
St. Christoph, Arlberg, Austria

Editors
H. Wachter · H. Ch. Curtius · W. Pfleiderer

Walter de Gruyter · Berlin · New York 1982

Editors
Prof. Dr. Helmut Wachter
Institute of Medical Chemistry and Biochemistry
University of Innsbruck
Fritz-Pregl-Strasse 3
A-6020 Innsbruck, Austria

Prof. Dr. Hans Christoph Curtius
Division of Clinical Chemistry
Department of Pediatrics
University of Zurich
Steinwiesstrasse 75
CH-8032 Zurich, Switzerland

Prof. Dr. Dr. h. c. Wolfgang Pfleiderer
Department of Chemistry
University of Konstanz
P.O.B. 5560
D-7730 Konstanz, F. R. of Germany

Library of Congress Cataloging in Publication Data

Winter Workshop on Pteridines
(1st : 1982 : Sankt Christoph am Arlberg, Austria)
Proceedings/First Winter Workshop on Pteridines,
March 7-10, 1982, St. Christoph, Arlberg, Austria.
(Biochemical and clinical aspects of pteridines; v. 1)
Bibliography: p.
Includes indexes.
Contents: v. 1. Cancer. Immunology. Metabolic diseases --
1. Pteridines-Congresses. 2. Pteridines-Diagnostic-use-Congresses.
I. Wachter, H. (Helmut), 1929-
II. Curtius, H. Ch. (Hans-Christoph) III. Pfleiderer,
Wolfgang, 1927- · IV. Title. V. Series
[DNLM: 1. Pteridines-Congresses. QD 401 B615 1982]
QP801.P69W56 1982 616.07 82-18338
ISBN 3-11-008984-X (U.S.)

CIP-Kurztitelaufnahme der Deutschen Bibliothek

Cancer, immunology, metabolic diseases:
proceedings/1. Winter Workshop on Pteridines, March 7-10, 1982.
Ed. H. Wachter ... - Berlin; New York: de Gruyter, 1982.
(Biochemical and clinical aspects of pteridines; Vol. 1)
ISBN 3-11-008984-X
NE: Wachter, Helmut [Hrsg.]; Winter Workshop on Pteridines
<01, 1982, Sankt Christoph>; GT

 Printing: Karl Gerike, Berlin. – Binding: Dieter Mikolai, Berlin. – Printed in Germany.

PREFACE

The "First Winter Workshop on Pteridines" held in St. Christoph, Arlberg (Austria) between March 7 - 10, 1982, centred around the developments and growing interest in clinical applications of pteridines.

The purpose of the Workshop was to bring together scientists in the fields of Biochemistry, Clinical Biochemistry, Analytical Chemistry and Immunology, as well as clinicians presently engaged in work on pteridines.

In addition to the biosynthesis and chemical analysis of naturally-occurring pteridines, the other major goal was to establish clinical implications in cancer detection, malignant cell growth control, immune response, metabolic diseases and neurological and psychiatric disorders.

Participants from France, West Germany, Switzerland and Austria reported on their most recent results achieved in connection with their research.

The contents of this book comprise specific reports covering six main topics:

- Fundamental chemistry for analysis and the biochemistry of naturally-occurring pteridines
- Quantitative chemical analysis of pteridines
- Pteridines in cancer detection and monitoring
- Pteridines in immunology
- Pteridines in proliferation and differentiation
- Pteridines in metabolic diseases

Data concerning pteridines as a diagnostic tool which may be used on neoplasias in haematology, gynaecology, paediatrics and in neurology as well as in psychiatry, are extensively described.

The clinical use of pteridines is important in the diagnosis and therapy of atypical phenylketonuria, the prediction of allograft rejections and in virology.

It is for this reason that the book provides information for the benefit of paediatricians, immunologists, virologists, oncologists, haematologists, gynaecologists, neurologists and psychiatrists, as well as all those involved in malignant cell growth control.

One chapter deals with the role of pteridines in the regulation of proliferation and differentiation and will thus attract the attention of biochemists.

The Editors hope this book will serve, not only as a source of reference for factual information, but will also afford any interested investigator an easy access to clinical research on pteridines.

We wish to thank de Gruyter Publishers · Berlin · New York for publishing the First Winter Workshop`s reports on pteridines.

H. Wachter
H.-Chr. Curtius
W. Pfleiderer

LIST OF PARTICIPANTS

Bichler, A.
Department of Obstetrics and Gynaecology, University of Innsbruck, Anichstrasse 35, A-6020 Innsbruck, Austria.

Blau, N.
Division of Clinical Chemistry, Department of Pediatrics, University of Zürich, Steinwiesstrasse 75, CH-8032 Zürich, Switzerland.

Curtius, H.-Chr.
Division of Clinical Chemistry, Department of Pediatrics, University of Zürich, Steinwiesstrasse 75, CH-8032 Zürich, Switzerland.

Dhondt, J.L.
Laboratory of Biochemistry, Faculty of Medicine, 56 rue du port, 59046 Lille Cédex, France.

Fink, M.
III. Clinic of Internal Medicine, University of Munich, Klinikum Großhadern, D-8000 Munich 70, F.R. of Germany.

Frater, M.
Department of Pediatrics, University of Zürich, Steinwiesstrasse 75, CH-8032 Zürich, Switzerland.

Fuchs, D.
Institute of Medical Chemistry and Biochemistry, University of Innsbruck, Fritz-Pregl-Strasse 3, A-6020 Innsbruck, Austria.

Ghisla, H.
Institute of Biochemistry, University of Konstanz, P.O.B. 5560, D-7750 Konstanz, F.R. of Germany.

Gröbner, P.
Institute of Medical Chemistry and Biochemistry, University of Innsbruck, Fritz-Pregl-Strasse 3, A-6020 Innsbruck, Austria.

Häusermann, M.
Division of Clinical Chemistry, Department of Pediatrics, University of Zürich, Steinwiesstrasse 75, CH-8032 Zürich, Switzerland.

Hausen, A.
Institute of Medical Chemistry and Biochemistry, University of Innsbruck, Fritz-Pregl-Strasse 3, A-6020 Innsbruck, Austria.

Hetzel, H.
Department of Obstetrics and Gynaecology, University of Innsbruck, Anichstrasse 35, A-6020 Innsbruck, Austria.

Huber, Ch.
Department of Internal Medicine, University of Innsbruck, Anichstrasse 35, A-6020 Innsbruck, Austria.

Körber, F.
Editor of the Journal of Clinical Chemistry and Clinical Biochemistry, Horazweg 31, D-1000 Berlin 42, F.R. of Germany.

Kostron-Krainz, Ch.
Department of Pediatrics, University of Innsbruck, Anichstrasse 35, A-6020 Innsbruck, Austria.

Leupold, D.
Department of Pediatrics, University of Ulm, Prittwitzstrasse 43, D-7900 Ulm/Donau, F.R. of Germany.

Loidl, P.
Institute of Medical Chemistry and Biochemistry, University of Innsbruck, Fritz-Pregl-Strasse 3, A-6020 Innsbruck, Austria.

Margreiter, R.
Department of Surgery I, University of Innsbruck, Anichstrasse 35, A-6020 Innsbruck, Austria.

Pfleiderer, W.
Department of Chemistry, University of Konstanz, P.O.B. 5560, D-7750 Konstanz, F.R. of Germany.

Reibnegger, G.
Institute of Medical Chemistry and Biochemistry, University of Innsbruck, Fritz-Pregl-Strasse 3, A-6020 Innsbruck, Austria.

Rembold, H.
Max-Planck-Institute for Biochemistry, D-8033 Martinsried, F.R. of Germany.

Rey, G.
Children`s Hospital for Metabolic Diseases, rue de sevres 149, F-75730 Paris, France.

Rokos, H.
Research Laboratory, Henning Berlin GmbH, Komturstrasse 19-20, D-1000 Berlin, F.R. of Germany.

Rokos, K.
Research Laboratory, Henning Berlin GmbH, Komturstrasse 19-20, D-1000 Berlin, F.R. of Germany.

Schweiger, M.
Institute of Biochemistry, University of Innsbruck, Innrain 52, A-6020 Innsbruck, Austria.

Schweiger, Monika
Institute of Biology and Genetics, University of Innsbruck, Schöpfstrasse 41, A-6020 Innsbruck, Austria.

Staudenmann, W.
Division of Clinical Chemistry, Department of Pediatrics, University of Zürich, Steinwiesstrasse 75, CH-8032 Zürich, Switzerland.

Wachter, H.
Institute of Medical Chemistry and Biochemistry, University of Innsbruck, Fritz-Pregl-Strasse 3, A-6020 Innsbruck, Austria.

Wachter, I.
Specialist in Children`s Diseases, Defreggerstrasse 8, A-6020 Innsbruck, Austria.

Weber, R.
de Gruyter Publishers, Director, Science Division, Genthiner Strasse 13, D-1000 Berlin 30, F.R. of Germany.

Ziegler, I.
Institute of Toxicology and Biochemistry of the Society of Radiation and Environmental Research, Dept. of Cell Chemistry, Arcisstrasse 16, D-8000 Munich, F.R. of Germany.

CONTENTS

1

FUNDAMENTAL CHEMISTRY FOR ANALYSIS AND THE BIOCHEMISTRY OF NATURALLY-OCCURRING PTERIDINES

PROPERTIES OF NATURALLY-OCCURRING PTERIDINES

Wolfgang Pfleiderer
Fakultät für Chemie, Universität Konstanz
Postfach 5560, D-7750 Konstanz

Introduction

The name "pteridine", which has its origin in the Greek word pteron = wing, was given in 1941 as a synonym to the pyrazino-[2,3-d]-pyrimidine ring system by Heinrich Wieland (1) accounting for the fact that the first derivatives of this heterocycle isolated from natural sources have been obtained from butterfly wings (2). These pigments and the majority of the naturally occurring pteridines turned out to be derivatives of 2-aminopteridin-4-one, of "pterin" which is modified in the 6- and/or 7-position by various substituents. A second, much smaller group of natural pteridine derivatives is derived from lumazine, which is the trivial name for pteridin-2,4-dione.

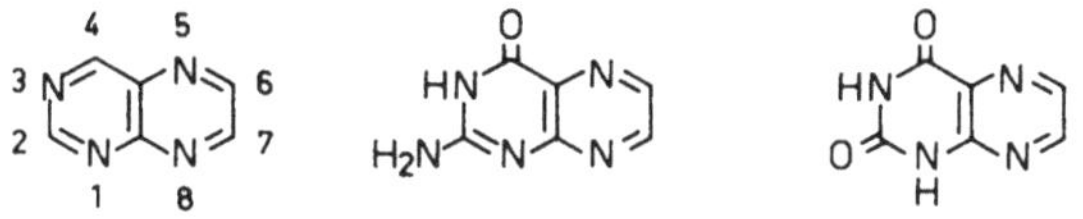

PTERIDINE PTERIN LUMAZINE

The natural occurrence of pterins and lumazines usually in small quantities requires a simple and very accurate method of characterization and identification for such molecules as offered in the UV-spectroscopy. Basic and acidic properties of these compounds have hereby been taken into account in order to compare corresponding molecular species such as monocations, neutral forms, monoanions, dianions, etc., which in general show characteristic differences in their spectral behaviour. Various

Biochemical and Clinical Aspects of Pteridines · Volume 1
Cancer · Immunology · Metabolic Diseases

types of pterin and lumazine derivatives will be presented and discussed regarding the structural and spectral relationship as a most important basis for future investigations of natural and synthetic pteridines.

Results

The most prominent structure of pterin in water is represented by the 2-amino-4-oxo-3,4-dihydro tautomer (3) indicating that an amino group is thermodynamically more stable than the imino dihydro form, whereas a hydroxy function adjacent to a ring nitrogen atom favours from analogous energetical reasons an amide and no iminol configuration.

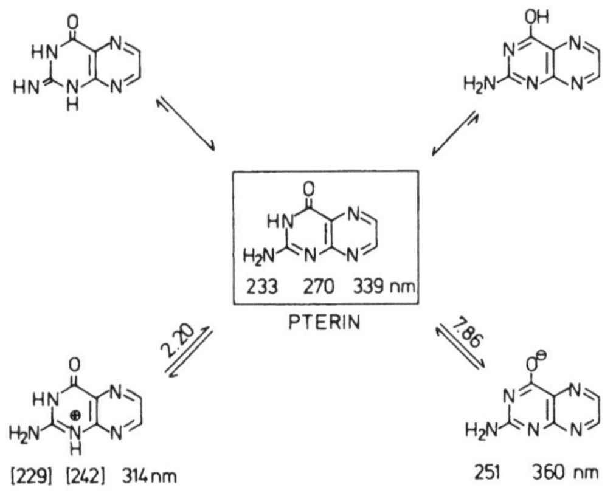

Pterin is an amphoteric substance in the normal pH range with a basic pK_a value of 2.20 and an acidic pK_a of 7.86. Comparisons of the UV-spectra of the neutral form with its cation and anion indicate that protonation causes a hypsochromic shift of the characteristic longwave length absorption band, whereas deprotonation to the monoanion is associated with a bathochromic shift. This spectral behaviour may be regarded as normal since protonation at N-1 localizes π-electrons in a amidine-type resonance and shortens this way the chromophoric system, which is becoming electron-richer in the anion, as revealed from the opposite effect (Fig. 1).

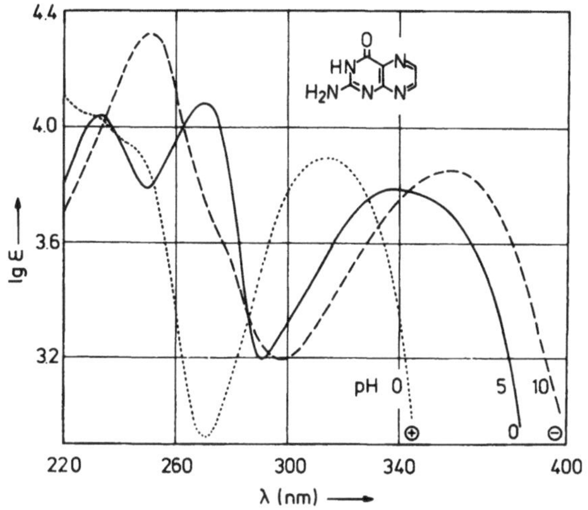

Fig. 1. UV-Spectra of the neutral, monoanion and cation species of pterin at different pH values.

Introduction of an aliphatic substituent in position 6 or 7 of the pterin nucleus does expectedly not alter the physical properties very much as seen from the biochemically most important biopterin.

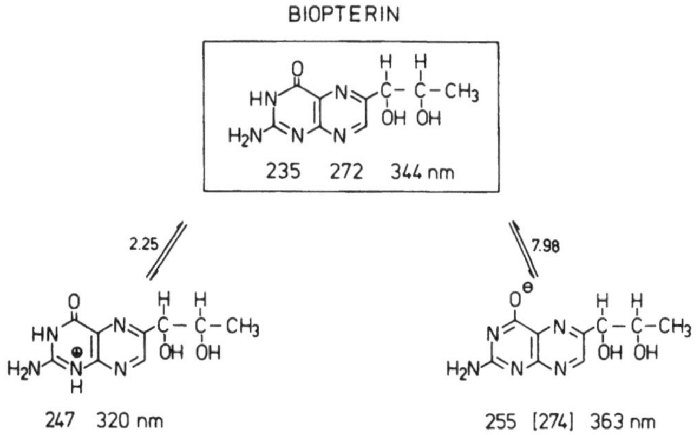

The dihydroxypropyl sidechain causes a small positive inductive effect shifting the UV-spectra of the various molecular species to slightly longer wavelengths (Fig. 2) in comparison to pterin and making the molecule a little bit more basic and less acidic respectively from the same reasons.

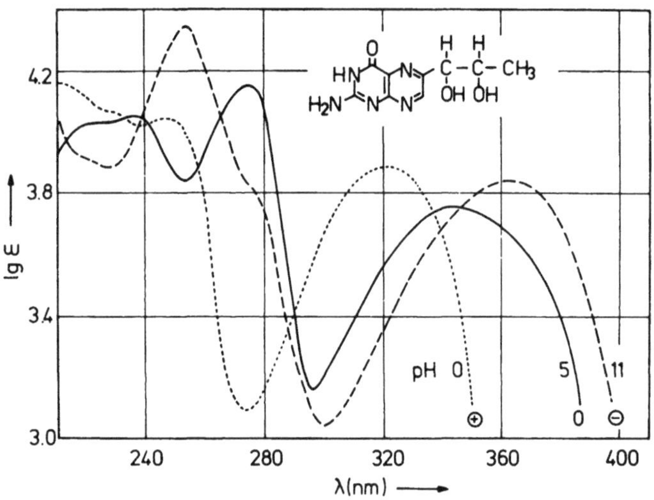

Fig. 2. UV-Spectra of the neutral, monoanion and cation species of biopterin.

Recently a new type of naturally-occurring pteridine has been isolated from Euglena gracilis (4) showing for the first time that in the pterin series the 2-amino group has been modified to a 2-dimethylamino group. Since this function is a much stronger electron-donor it is obvious that the UV-absorption is shifted to longer wavelengths. The neutral form of euglenapterin reveals a maximum at 352 nm and a broad shoulder at 376 nm.

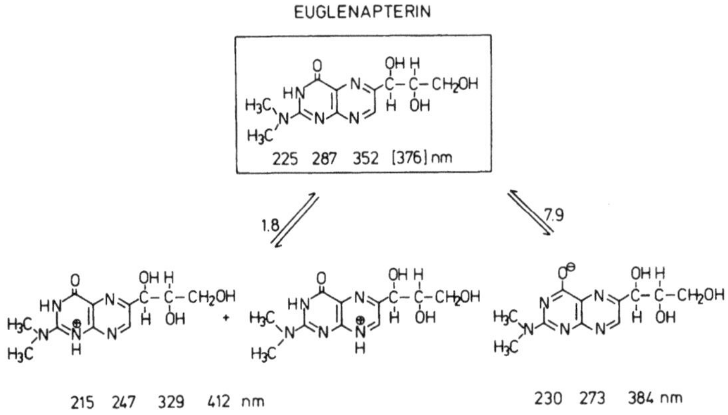

The monoanion shows the expected properties, but a more complex situation has to be considered for the cation, for which a pK_a of 1.8 has been determined. The decreased basicity of euglenapterin is due to a steric hindrance effect of the 2-dimethylamino group on the adjacent N-1 atom so that protonation does not take place uniformly at this position but also at the vinylogous N-8 atom. This new cation species possesses a long cross-conjugated π-electron system shifting the long wavelength band even to the visible region. The presence of a mixture of two cation species of euglenapterin at pH -1 is clearly be seen from the UV-spectrum, in which the maximum at 329 nm is attributed to the 1-H and the peak at 412 nm to the 8-H cation form (Fig. 3). An unusual behaviour of this type is of great analytical value for structural assignment and elucidations.

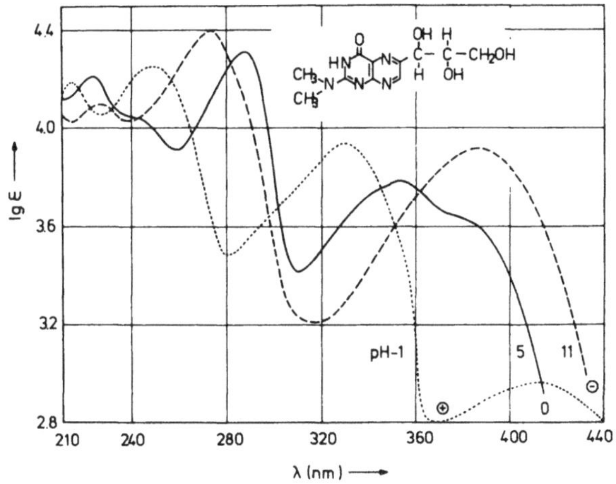

Fig. 3. UV-Spectra of the neutral, monoanion and cation species of euglenapterin.

The formal oxidation of pterin in position 7 to isoxanthopterin (5) creates a molecular form which is characterized by the built-in linear chromophor terminated by the 2-amino group as the donor and the 7-oxo function as the acceptor. This conjugated amide stabilizes the presented tautomeric lactam form and is mainly responsible for the observed molecular properties.

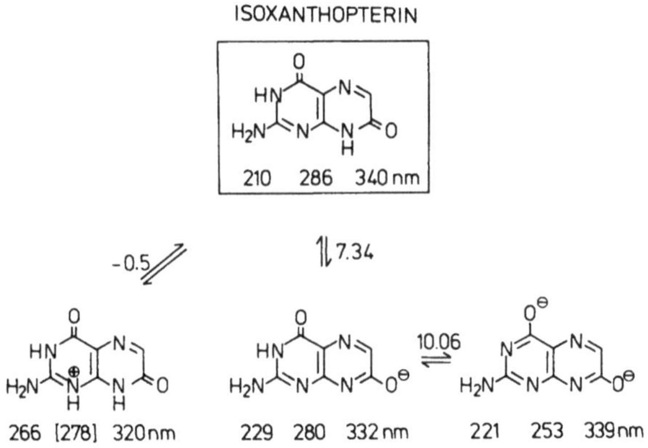

The amide resonance explains nicely why isoxanthopterin is a very weak base with apK_a of - 0.5, since protonation at N-1 is equivalent to a shortening of the mesomeric system noticed also from the hypsochromic shift of the long wavelength absorption band in the cation. Another interesting and characteristic feature of isoxanthopterin is associated with the monoanion formation showing also a hypsochromic shift compared to the neutral form. This is explained by the formation of a new chromophoric system in the monoanion enclosing the carbonyl group in position

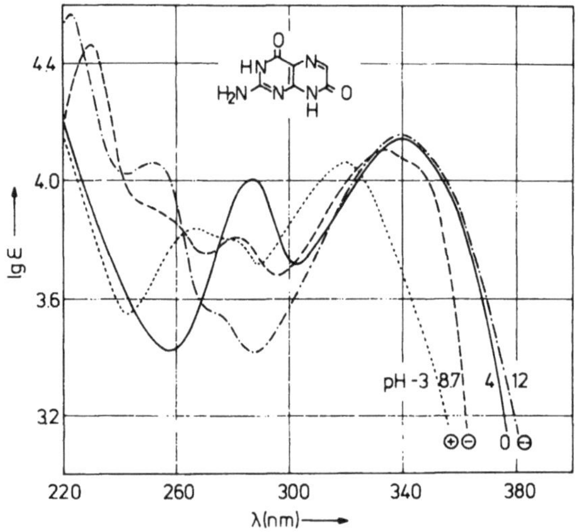

Fig. 4. UV-Spectra of the neutral, monoanion, dianion and cation species of isoxanthopterin.

4 and the 7-hydroxide function. In the dianion finally the normal bathochromicity is observed again (Fig. 4).

In the yellow butterfly pigment xanthopterin (6) causes the amide function in the pyrazine moiety a strong perturbation of the pterin chromophor effecting especially the ground-state energy level which is increased and allows therefore an easier excitation, which is equivalent with the observed red-shift towards the visible region.

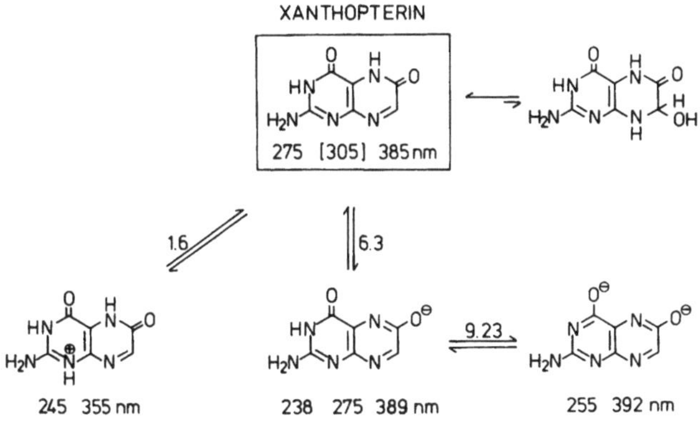

The neutral form of xanthopterin bears another complication in its structure since it is in solution an equilibrium mixture between the normal and the covalently hydrated species. The latter molecular form is recognized from the UV-spectrum taken at pH 4 firstly by the relatively low extinction of the long wavelength band and secondly the shoulder at 305 nm, which coincides with the absorption maximum of the 7,8-dihydroxanthopterin (7), the structural model of xanthopterin-hydrate. The mono- and dianions of xanthopterin behave normally and consist of anhydrous species (Fig. 5).

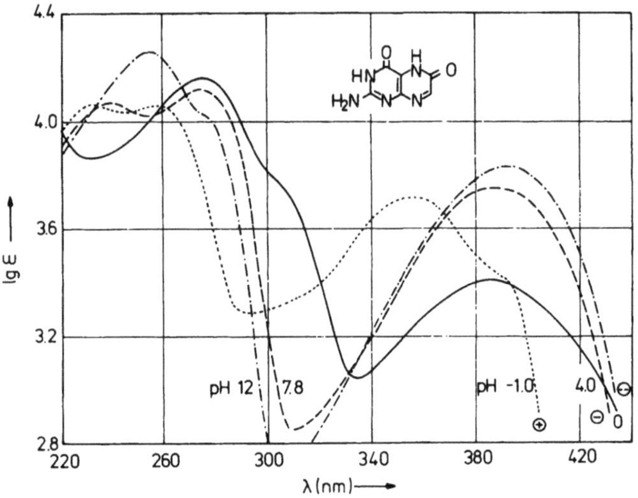

Fig. 5. UV-Spectra of the neutral, monoanion, dianion and cation species of xanthopterin.

Leucopterin, the colourless pigment of the cabbage butterfly, reveals another characteristic UV-spectrum of the neutral species, whereby the whole shape of the curve is more informative than the sites of the absorption maxima alone (8).

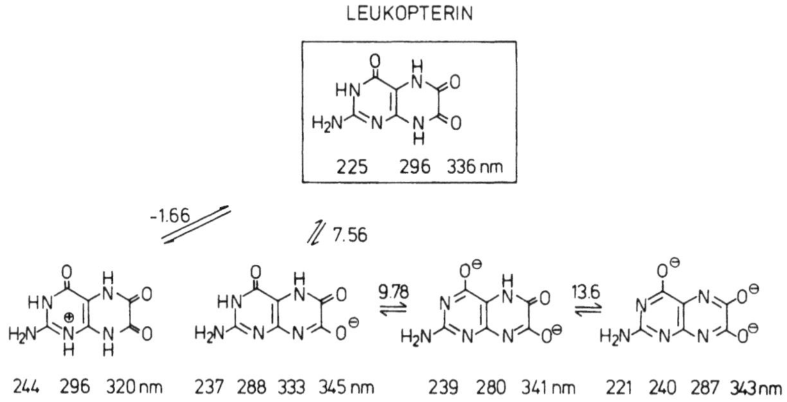

The monoanion species taken at pH 8.7 cannot be a pure form due to the overlapping acidic pK_a of 7.56 and 9.78. Furthermore the appearance of a double maximum at 333 and 345 nm in the monodeprotonated form may account for the presence of more than one monoanionic species as for example deprotonation from N-8, N-3 or N-5 respectively. Further details studied in this area, however, have to be achieved to prove this fact. Protonation of leucopterin proceeds only in super-acidic medium due to the very weak basic properties of this molecule (Fig. 6).

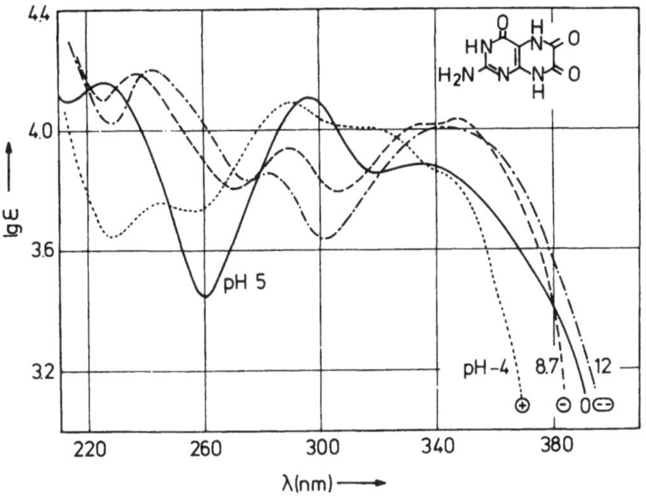

Fig. 6. UV-Spectra of the neutral, monoanion, dianion, and cation species of leucopterin.

The chemistry of hydrogenated pterins plays an important role in nature due to the redox behaviour of various dihydro- and tetrahydro derivatives in biological systems. Unfortunately the relative instability and tendency to autoxidation of these substances has not encouraged to a broad systematic investigation of this field so far (9). Considering pterin as the basic molecule of most naturally-occurring pteridines there is the possibility of reduction to one tetrahydro form involving the 5,6- and 7,8-C=N double bonds and actually five structural isomers for the dihydro form that is the 5,6, 5,8, 7,8, 6,7 and 6,7,8-dihydropterin.

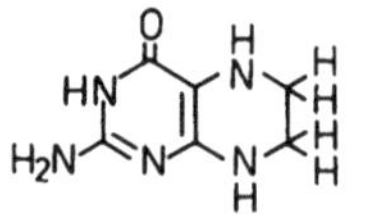

5.6.7.8-TETRAHYDRO-PTERIN

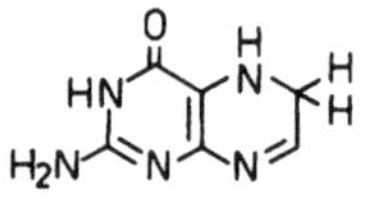

5.6-DIHYDRO-

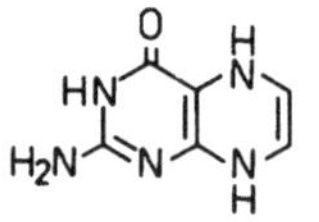

5.8-DIHYDRO-

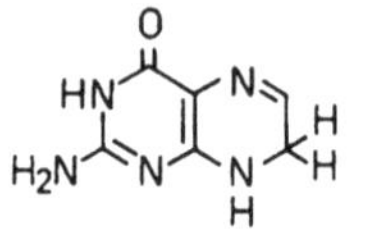

7.8-DIHYDRO-

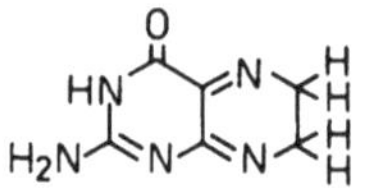

6.7-DIHYDRO-

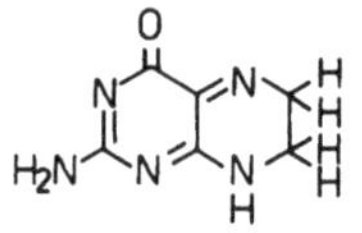

6.7.8-DIHYDRO-

Other isomers accounting for different tautomeric forms are less probable and have not to be considered from energetical reasons analogously to the heteroaromatic counterparts. From the above mentioned five dihydropterins only the 7,8-dihydro derivative has so far been synthesized, isolated, and characterized spectrophotometrically (10). This type is expected to be the thermodynamically most stable form at all, since the 8-NH function is part of a vinylogous amide resonance stabilization involving the carbonyl group at 4 position. 6-Methyl-7,8-dihydropterin (11), which is a simple model for the natural analogs, 7,8-dihydrobiopterin and -neopterin respectively reveal the expected, up-shifted more basic and less acidic properties due to the partial hydrogenation and the loss of the heteroaromatic character of the pyrazine ring.

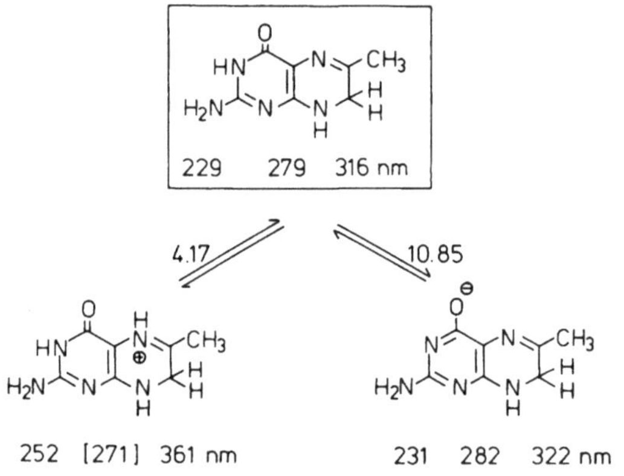

Besides these normal changes a striking phenomenon turned up on protonation leading to an anormalous bathochromic shift of the long wavelength absorption band, which seems to be very characteristic for this type of compounds (Fig. 7). It has been proven that protonation does take place in these cases at N-5 and not at N-1, which would give raise to a hypsochromic change in the UV-spectrum. Anion formation, on the other hand, does not effect the spectrophotometric properties very much at all and can be considered as a normal red-shift on a more inductive basis.

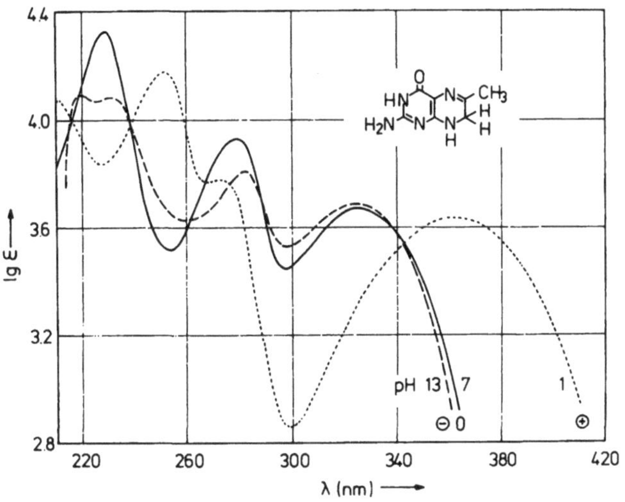

Fig. 7. UV-Spectra of the neutral, monoanion, and cation species of 6-methyl-7,8-dihydropterin.

Analogously 7,8-dihydro-D-neopterin shows the same spectral features indicating a small electronic interaction of a saturated side-chain with the chromophoric system (Fig. 8).

On the other hand there is a dramatic spectral change of the 7,8-dihydropterin chromophor if a carbonyl group is located in the 1'-position of the side-chain as in the yellow Drosophila eye-pigments sepiapterin (11) and deoxysepiapterin. The strong absorption band in the visible region is due to a mesomeric interaction of the 2-amino group as the electron-donor and the

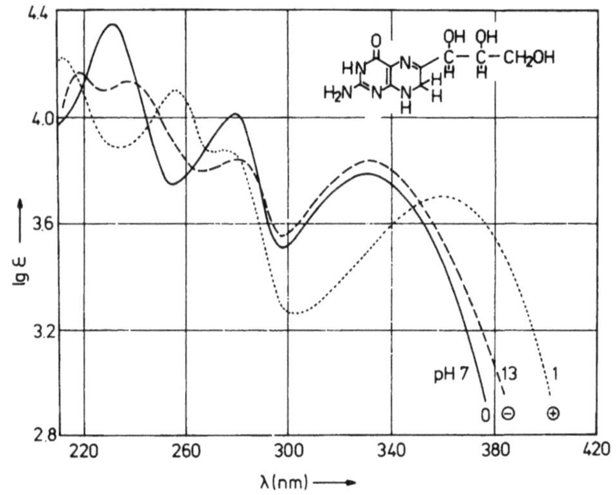

Fig. 8. UV-Spectra of the neutral, monoanion and cation species of 7,8-dihydro-D-neopterin.

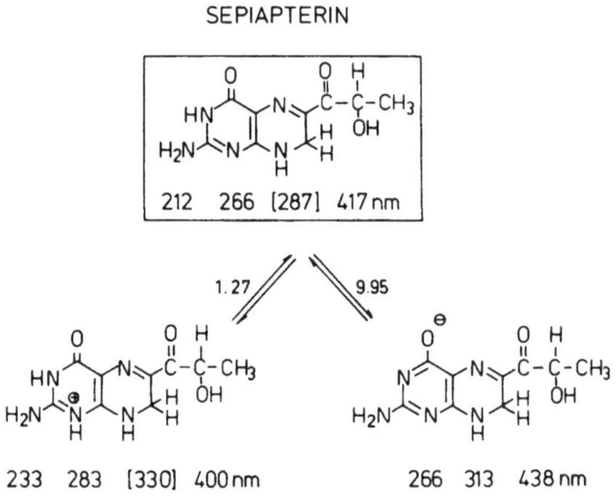

keto function in the side-chain as the electron-attracting group creating again a linear merocyanine-type chromophor. The partial reduction in 7,8-position favours this interaction to a large extent by interrupting the cyclic heteroaromatic resonance of the pyrazine moiety. From the hypsochromic shift in the cation can be concluded that protonation takes place at N-1 shortening the chromophoric system by one conjugated double-bond. Anion formation, however, causes a bathchromicity of the same extent of magnitude suggesting an elongation of the chromophor by one additional conjugated double-bond (Fig.9).

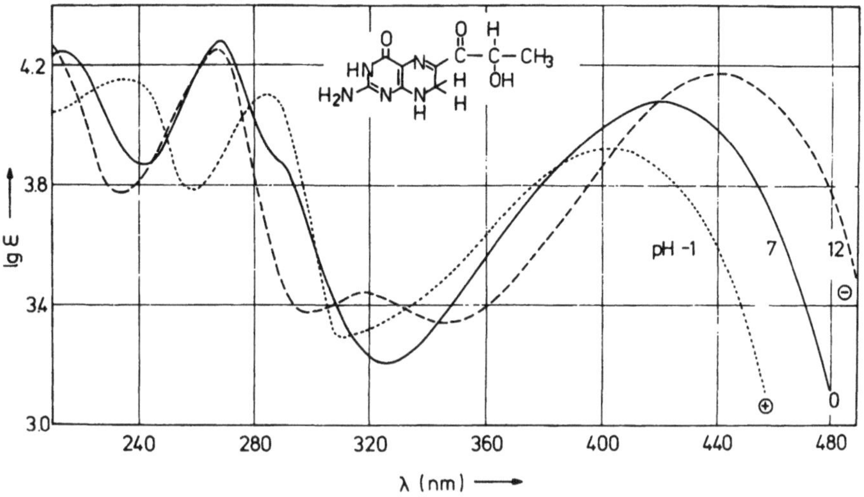

Fig. 9. UV-Spectra of the neutral, monoanion and cation species of sepiapterin.

The physical properties of deoxysepiapterin, formerly called isosepiapterin, are expectedly very similar, whereby the small spectral differences to sepiapterin reside in the higher saturation of the side-chain.

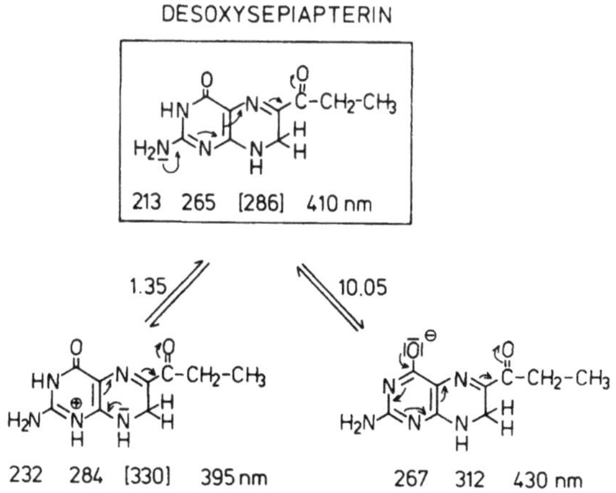

Considerations of the other dihydropterins reveal that electrochemical reductions of simply substituted pterins (12) take place at the 1,4-diazabutadiene portion of the pyrazine ring forming initially a short-lived 5,8-dihydropterin structure, which tautomerizes in an acid-catalysed reaction to the 7,8-dihydro isomer or oxidizes under aerobic conditions back to the starting material. 5,8-Dihydropterins are formal 8π-electron-antiaromatic heterocycles, which cannot be expected to be stable enough for isolation. Introduction of an electron-attracting substituent at position 6 or 7 can stabilize a 5,8-dihydro structure by resonance at least under anaerobic conditions as demonstrated in the photochemical degradation of biopterin to 5,8-dihydropterin-6-carbaldehyde (13). This unusual structure is reflected in its spectrum with the characteristic long wavelength absorption of low extinction at 475 nm.

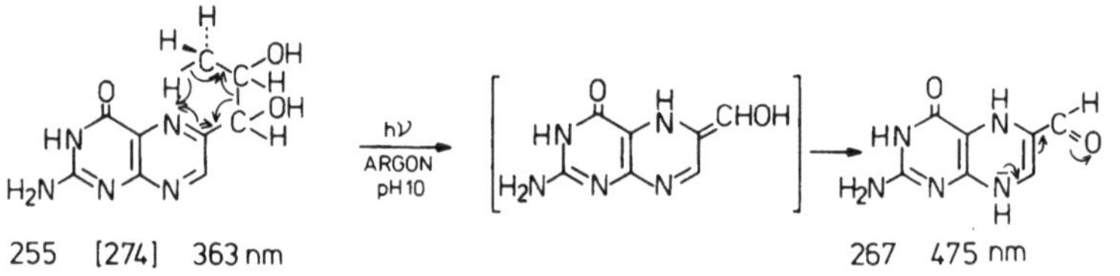

The photochemical irradiation of the isomeric 7-(1',2'-dihydroxypropyl)-pterin proceeds analogously with formation of 5,8-dihydropterin-7-carbaldehyde, which absorbs even at 500 nm but is also oxidized immediately by air.

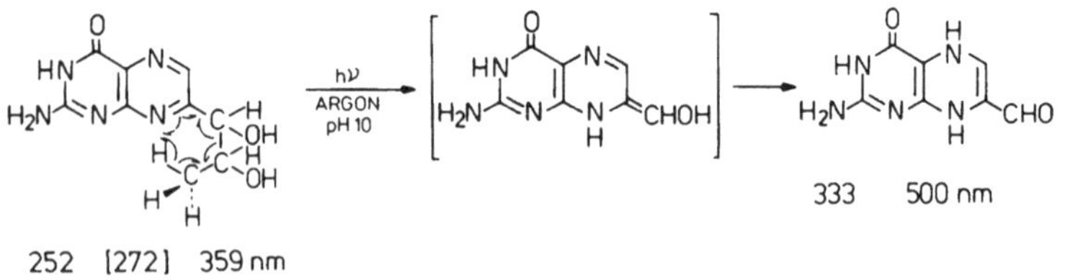

Some physical and chemical properties of the labile quinonoid 6,7,8-dihydropterins are also knwon from the redox shuttle of the 5,6,7,8-tetrahydropterins in the enzymatic hydroxylation processes converting phenylalanine into tyrosine (14,15). More chemically orientated investigations (16) localize the intermediary tautomeric quinonoid structure energetically between the 5,6,7,8-tetrahydro- and the 7,8-dihydro form and associate with this labile molecular form an UV-spectrum with an absorption maximum at 303 nm at pH 8. The same spectral properties apply also to other 6-substituted 6,7,8-dihydropterin derivatives, if the side-chain is saturated and there are not further electronic interactions. Successful isolation of such species in solid state is very much unlikely due to their thermodynamical instability.

Further reduction of pterins leads to 5,6,7,8-tetrahydropterins, which can be regarded as 5,6-diaminopyrimidine analogs having interconnected the two ortho-situated amino groups by a saturated C_2-bridge. 5,6,7,8-Tetrahydropterin (19) and its 6-substituted derivatives show two basic (1.3; 5.6) and one acidic pK_a value (10.6) in the normal pH region, whereby the stronger basic properties are reflected by the first basic pK_a of 5.6. Monocation formation takes place at N-5 according to the hypsochromic shift of the long wavelength absorption in the neutral form. The second basic pK_a of 1.3 is describing the monocation - dication equilibrium with the second protonation at N-1 obviously. It should also be noticed that in this series anion formation is again associated with a small blue-shift in the UV-absorption. Furthermore does the autoxidation of tetrahydropterins increase with raising pH due to the greater availability of electrons.

5,6,7,8-Tetrahydrobiopterin, the natural cofactor in biological hydroxylation processes (14,15,17,18) reveal the same properties as seen from the following equilibria:

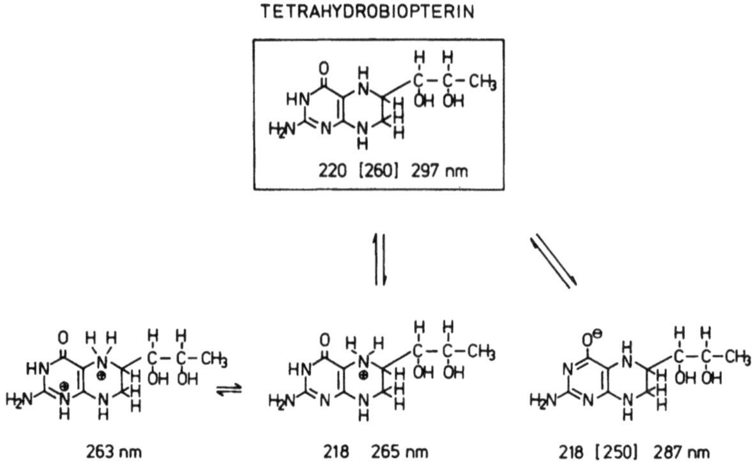

The UV-spectra are also in agreement with the more simple model substance 5,6,7,8-tetrahydropterin.

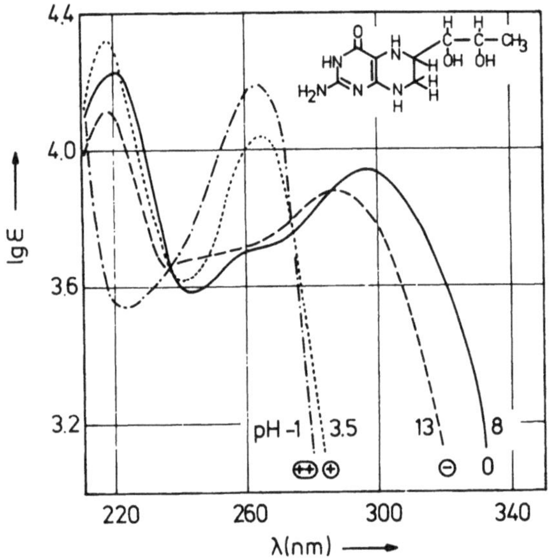

Fig. 10. UV-Spectra of the neutral, monocation, dication, and monoanion species of 5,6,7,8-tetrahydrobiopterin.

The pteridine nucleus may also be part of more complex molecules as recently realized during the structure elucidation of the drosopterins (20). The pentacyclic heterocyclic ring system is derived from 5,6,7,8-tetrahydropterin but does not have anything in common with this simple molecule due to the unsaturation of the annelated pyrrolo-, diazepino- and pyrimidine ring with a long wavelength absorbing merocyanine-type chromophor. Droso- and its enantiomer isodrosopterin absorb around 460-500 nm, depending on the various molecular species characterized by three basic and two acidic pK_a values. The complex equilibrium situations are described as follows:

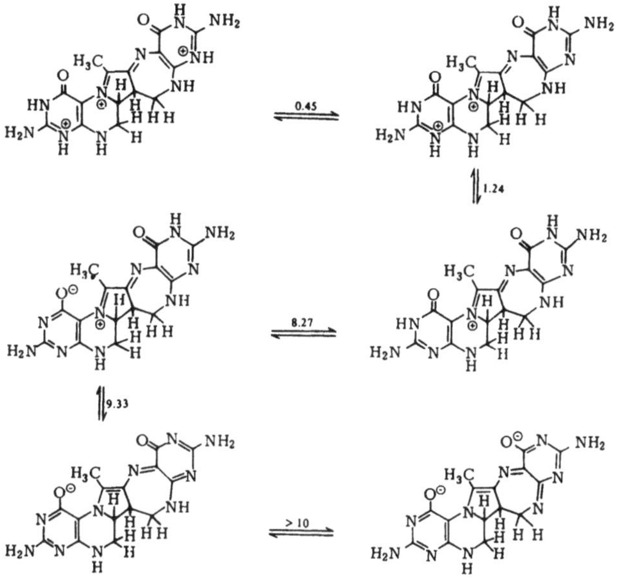

Besides the UV-absorption spectra (Fig. 11) additional information about the structures of natural products can arise from the determination of their chiroptical properties.

Droso- and isodrosopterin impress by their ORD-(Fig. 12) and CD-spectra (Fig. 13) (21) revealing a high molecular chirality associated with enormous rotational strengths of the various transitions.

In the lumazine series analogous considerations of the structural relationships based on chemical and physical properties have resulted in a good understanding of the molecular features. Lumazine itself (22) reveals in the normal pH range only acidic properties due to the amide functions in the pyrimidine moiety.

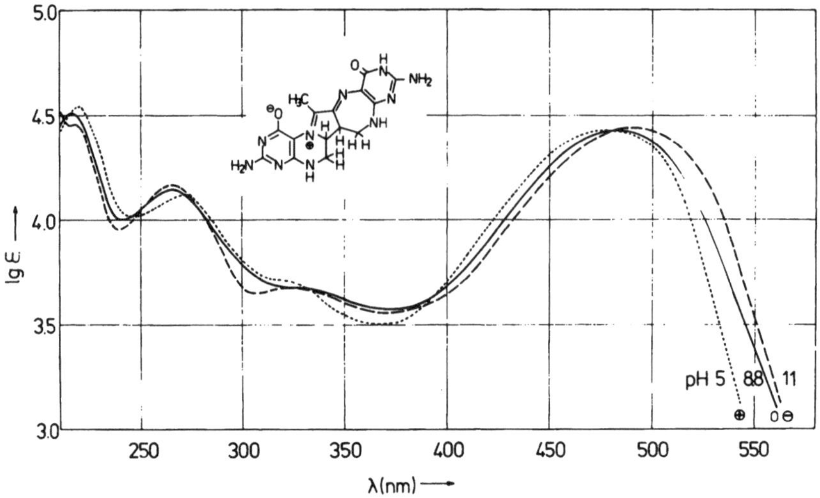

Fig. 11. UV-Spectra of the neutral (zwitterionic), monoanion, and monocation species of droso- and isodrosopterin.

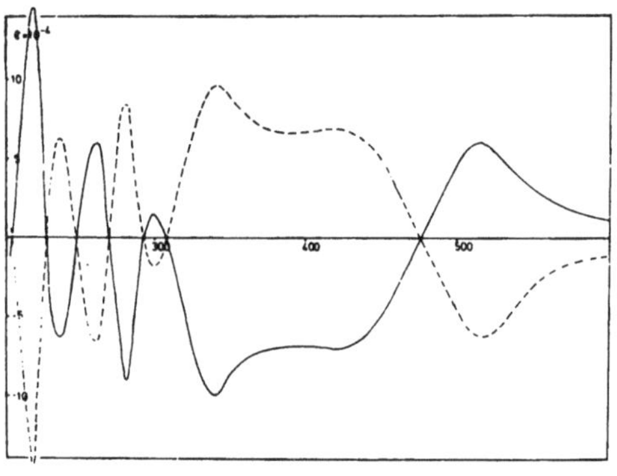

Fig. 12. ORD-Spectra of droso- and isodrosopterin in water.

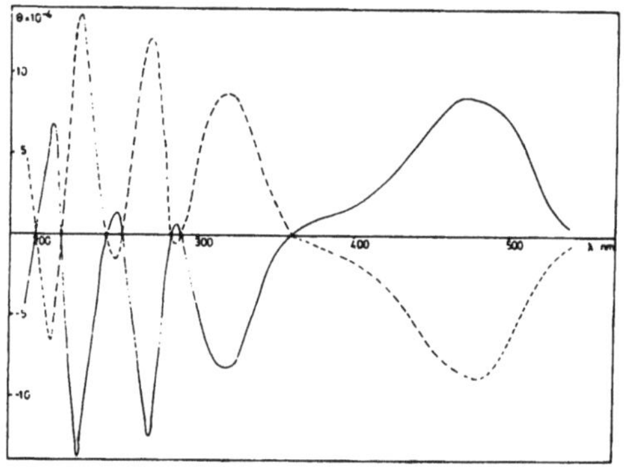

Fig. 13. CD-Spectra of droso- and isodrosopterin in water.

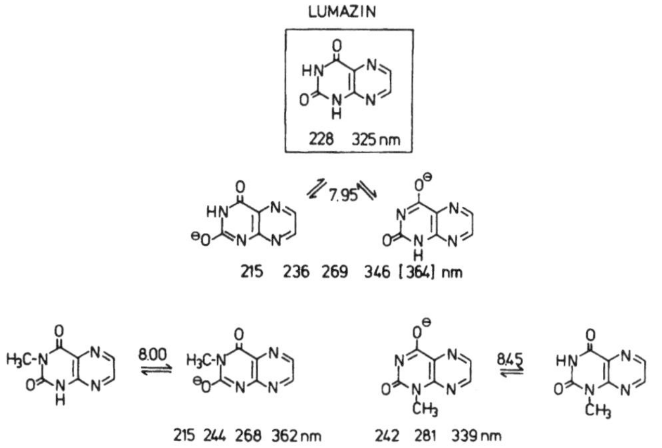

Some anomaly of this molecule arises on monoanion formation, which is actually a mixture of the N-1 and N-3 deprotonated species. This fact can be derived from a comparison of the first acidic pK_a of lumazine (7.95) with those of the 1-methyl- (8.45) and 3-methyllumazine (8.00) respectively. The closely related values indicate very similar acidities of both centers with only a slight preference for N-1 deprotonation. The monoanion spectrum of lumazine (pH 10.5) gives further proof for

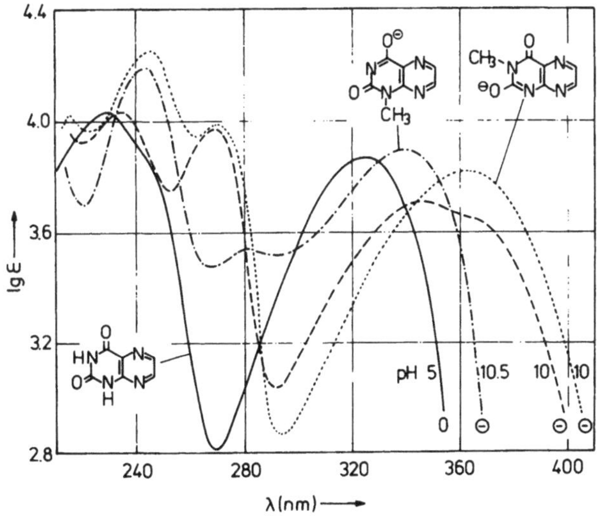

Fig. 14. UV-Spectra of neutral and monoanion species of lumazine and the anions of 1- and 3-methyllumazine.

the two species mixture by the absorption maximum at 346 nm corresponding to the N-3 deprotonated form and the shoulder at 364 nm, indicating the presence of the N-1 deprotonated species (Fig. 14).

N-8 Alkylation of lumazine gives rise to a new long wavelength absorbing chromophor characterized by a cross-conjugated π-electron system (23,24) involving the N-8 alkyl function as the donor and the carbonyl groups as the acceptor sites in the molecule. The 6,7,8-trimethyllumazine reveals a strong absorption band at 403 nm which is blue-shifted on protonation to the cation species (Fig. 15).

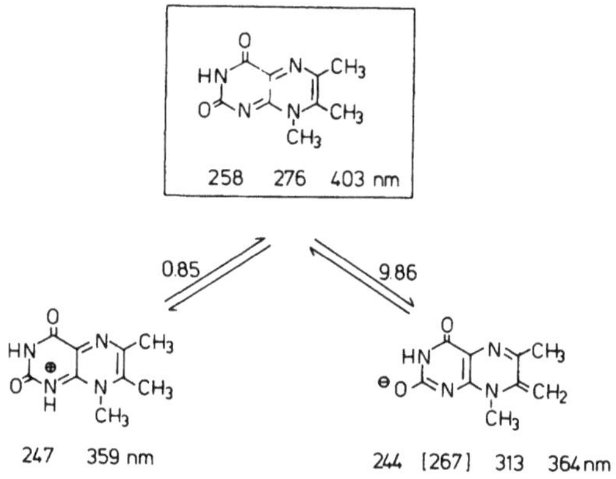

Monoanion formation is initiated by deprotonation from the 7-methyl group producing by resonance the more likely 6,8-dimethyl-7-methylenelumazine-anion species (25) of lower extinction and shorter wavelength absorption.

The first lumazine derivative found in nature and isolated from the yeast Eremothecium ashbyii (26,27) and ashbya gossipyii (28,29) respectively could be identified as 6,7-dimethyl-8-D-ribityllumazine. It is a direct precursor of riboflavin in the biosynthetic pathway (30).

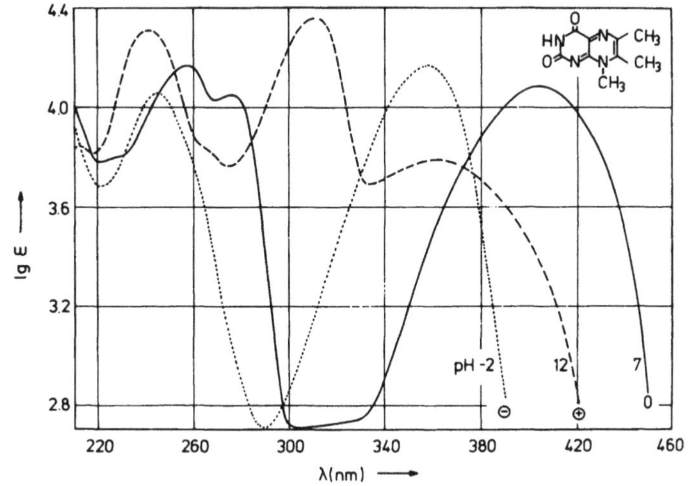

Fig. 15. UV-Spectra of the neutral, monoanion, and cation species of 6,7,8-trimethyllumazine.

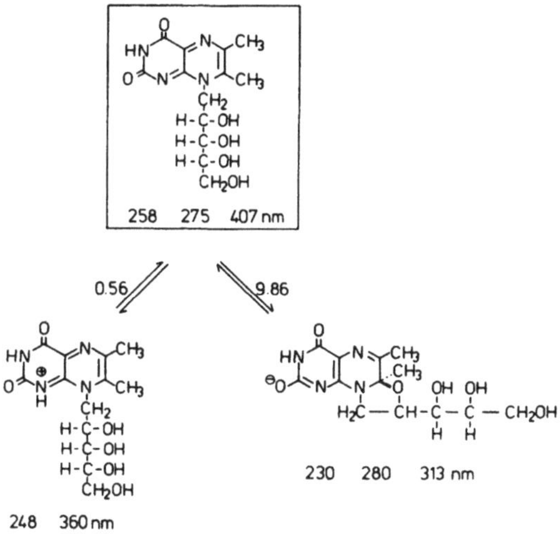

The yellow-coloured, yellow-green fluorescing 6,7-dimethyl-8-D-ribityllumazine reveals in its neutral form a characteristic UV/VIS-spectrum with a strong absorption band at 407 nm and a broad deep minimum between 300 and 340 nm. Protonation effects a hypsochromic shift to 360 nm as expected from N-1 substitution, whereas the anion spectrum with its longest absorption maximum at only 313 nm does not coincide with that of the

6,7,8-trimethyllumazine analogne.The spectral type resembles more a 7,8-dihydrolumazine derivative assuming that the mono-anion is actually derived from a nucleophilic attack of the ribityl side-chain onto position 7 forming a covalent intra-molecular adduct.

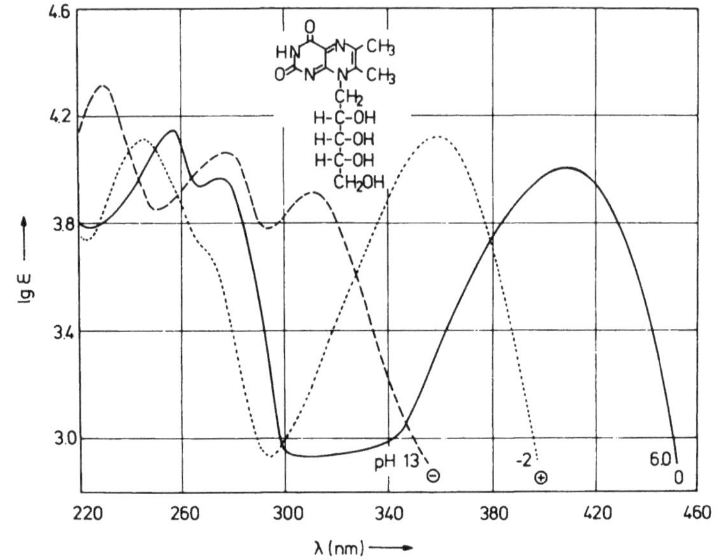

Fig. 16. UV-Spectra of the neutral, monoanion, and cation species of 6,7-dimethyl-8-D-ribityllumazine.

These examples illustrate nicely how different and versatile small structural changes may be reflected in the physical and chemical properties of such molecules.

Little information is available about dihydrolumazines, where-as the structural counterpart of the 5,6,7,8-tetrahydro deri-vative has to be seen again in the 5,6-diaminouracil (31).

Acylation and protonation at N-5 position localizes the donor electrons of the most basic center at this site and is regis-tered by a more or less strong hypsochromic shift of the long wavelength absorption band, whereas anion formation does not

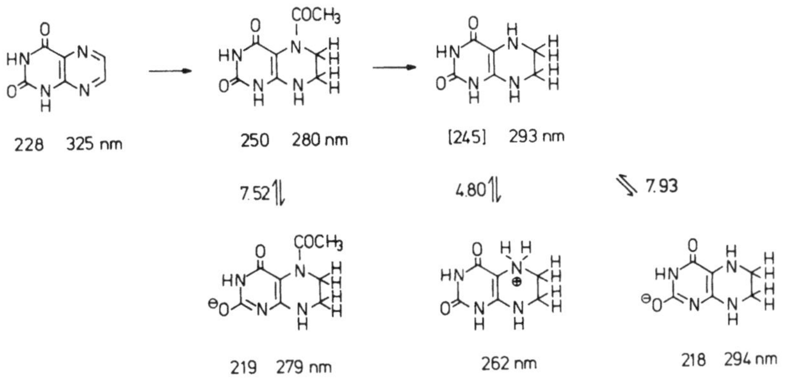

alter the shape of the UV-spectrum very much.

Finally it can be concluded from all these investigations that UV-spectral comparisons on the basis of pH-dependent pure molecular species arean important and very valuable tool in structural elucidations of new naturally-occurring pteridines in general.

References

1. Schöpf, C., Reichert, R., Riefstahl, K.: Liebigs Ann. Chem. 548, 83 (1941).
2) Pfleiderer, W.: Angew.Chem., Int.Ed.Engl. 3, 114 (1964).
3) Pfleiderer, W., Liedek, E., Lohrmann, R., Rukwied, M.: Chem.Ber. 93, 2015 (1960).
4) Böhme, M., Pfleiderer, W., Elstner, E.F., Richter, W.J.: Angew.Chem. 92, 474 (1980).
5) Pfleiderer, W., Rukwied, M.: Chem.Ber. 94, 1 (1961).
6) Pfleiderer, W., Liedek, E., Rukwied, M.: Chem.Ber. 95, 755 (1962).
7) Pfleiderer, W.: Chem.Ber. 107, 785 (1974).
8) Pfleiderer, W., Rukwied, M.: Chem.Ber. 94, 118 (1961).
9) Pfleiderer, W.: J.Inher.Metab.Dis. 1, 54 (1978).

10) Stuart, A., Wood, H.C.S., Duncan, D.: J.Chem.Soc.(C) 1966, 285.

11) Pfleiderer, W.: Chem.Ber. 112, 2750 (1979).

12) Lund, H., in "Chemistry and Biology of Pteridines", ed. Pfleiderer, W., W. de Gruyter, Berlin, 1975, p. 753.

13) Mengel, R., Pfleiderer, W., Knappe, W.R.: Tetrahedron Lett. 1977, 2817.

14) Kaufman, S., in "Advances of Enzymology", ed. Meister, A., John Wiley & Sons, New York, 1971, p.245.

15) Kaufman, S. in "Chemistry and Biology of Pteridines", ed. Pfleiderer, W., W. de Gruyter, Berlin, 1975, p. 291.

16) Archer, M.C., Scrimgeour, K.G.: Can.J.Biochem. 48, 278 (1970).

17) Levitt, M., Spector, S., Sjoerdsma, A., Udenfriend, S.: J.Pharmacol.Exp.Ther. 148, 1 (1965).

18) Hosoda, S., Glick, D.: J.Biol.Chem. 247, 4549 (1966).

19) Viscontini, M., Weilenmann, H.R.: Helv.Chim. Acta 41, 2170 (1958).

20) Theobald, N., Pfleiderer, W.: Chem.Ber. 111, 3385 (1978).

21) Schlobach, H., Pfleiderer, W.: Helv.Chim. Acta 55, 2525 (1972).

22) Pfleiderer, W.: Chem.Ber. 90, 2582 (1957).

23) Pfleiderer, W., Nübel, G.: Chem.Ber. 93, 1406 (1960).

24) Pfleiderer, W., Bunting, J.W., Perrin, D.D., Nübel, G.: Chem.Ber. 99, 3503 (1966).

25) Pfleiderer, W., Mengel, R., Hemmerich, P.: Chem.Ber. 104, 2273 (1971).

26) Masuda, T.: Pharm.Bull. Japan 4, 71; 72 (1956).

27) Masuda, T., Kishi, T., Asai, M.: Pharm.Bull. Japan 5, 598 (1957).

28) Plaut, G.W.E., Maley, G.F.: Arch.Biochem.Biophys. 80, 219 (1959).

29) Maley, G.F., Plaut, G.W.E.: J.Biol.Chem. 234, 641 (1959).

30) Harvey, R.A., Plaut, G.W.E.: J.Biol.Chem. 241, 2120 (1966).

31) Pfleiderer, W.: Liebigs Ann.Chem. 747, 111 (1971).

TETRAHYDROBIOPTERIN BIOSYNTHESIS

Hans-Christoph Curtius, Max Häusermann, Alois Niederwieser
Division of Clinical Chemistry, Department of Pediatrics, University of Zurich, CH-8032 Zurich, Switzerland

Sandro Ghisla
Department of Biology, University of Constance, Germany

Introduction

Our laboratory has been interested in phenylketonuria (PKU) for several years, especially in light of recent demonstrations by our own and other laboratories that certain atypical forms of PKU are caused by defects in tetrahydrobiopterin (BH_4) metabolism (1,2,3,4,5,6,7,8). Approximately 1-3% of PKU cases are caused by a lack of BH_4 due to an inborn error of metabolism. It is well known that BH_4 plays an important role in regulating tyrosine and tryptophan hydroxylase in the synthesis of the biogenic amine neurotransmitters, dopamine, norepinephrine, epinephrine, and serotonin (9,10). Therefore, many atypical PKU patients suffer from a lack of proper biogenic amine synthesis in the central nervous system. These patients develop a progressive neurological illness that is unresponsive to a low phenylalanine diet. The majority of atypical PKU patients die at an early age.

We have investigated several patients that are unable to synthesize dihydrobiopterin (BH_2) as well as patients that are deficient in quinonoid dihydropteridine reductase (DHPR) activity (11,12). Recently, we have discovered one patient with a diminished GTP-cyclohydrolase activity (5). Several of these atypical PKU patients are responsive to BH_4 administration alone (12), and

Biochemical and Clinical Aspects of Pteridines · Volume 1
Cancer · Immunology · Metabolic Diseases

others are responsive to BH_4 with supplemental neurotransmitter precursor therapy. Of course, a better knowledge of BH_4 biosynthesis in man would aid efforts to understand and treat these diseases. Our recent observation that BH_4 administration may be beneficial in certain cases of Parkinson's disease and endogenous depression (13) emphasizes the need for a further understanding of BH_4 biosynthesis and metabolism. Although our results (13,14) as well as those of others (15,16) on BH_4 administration in Parkinson's disease are only preliminary and require further investigation, they stress the importance of basic pterin research. It is with these ideas in mind that we have continued to explore BH_4 biosynthesis in certain mammals, including man.

Historical Perspectives on BH_4 Biosynthesis

Invertebrate systems

A great deal of the early investigations on BH_4 biosynthesis was accomplished on non-mammalian systems. In E.coli (17,18) and other invertebrates (19,20,21,22,23), an enzyme system was isolated that was not Mg^{++}-dependent and converted guanosine triphosphate (GTP) directly to D-erythro-dihydroneopterin triphosphate (NH_2TP) this enzyme was named GTP-cyclohydrolase I. These observations led to investigations by Krivi and Brown on the transformation of NH_2TP in BH_4 biosynthesis (see figure 1). According to their studies in *Drosophila melanogaster* (24), sepiapterin synthetase, the enzyme system responsible for the synthesis of sepiapterin from NH_2TP, consisted of 2 components referred to as enzyme A (M.W.=82,000) and enzyme B (M.W.=36,000). NADPH and a divalent cation (Mg^{++} was the most effective) were required for activity that was shown to be optimal at pH 7.4, and 30°C. The K_M for NH_2TP was 10 µM and a number of unconjugated pterins, including biopterin and sepiapterin, were inhibitors of the reaction. Dihydroneopterin (NH_2) would not serve as substrate in place of NH_2TP. Evidence was presented in support of a proposed reaction mechanism for the enzymatic conversion of NH_2TP to sepiapterin in which enzyme A catalyzes the production of a labile intermediate by nonhydrolic elimination of the phosphates of NH_2TP. According to this proposal, enzyme B catalyzed the conversion of this intermediate to sepiapterin in the presence of NADPH. These biosynthetic reactions are shown in figure 1. It should be noted that a similar reaction scheme was proposed by Viscontini and coworkers as early as 1973 (38,39).

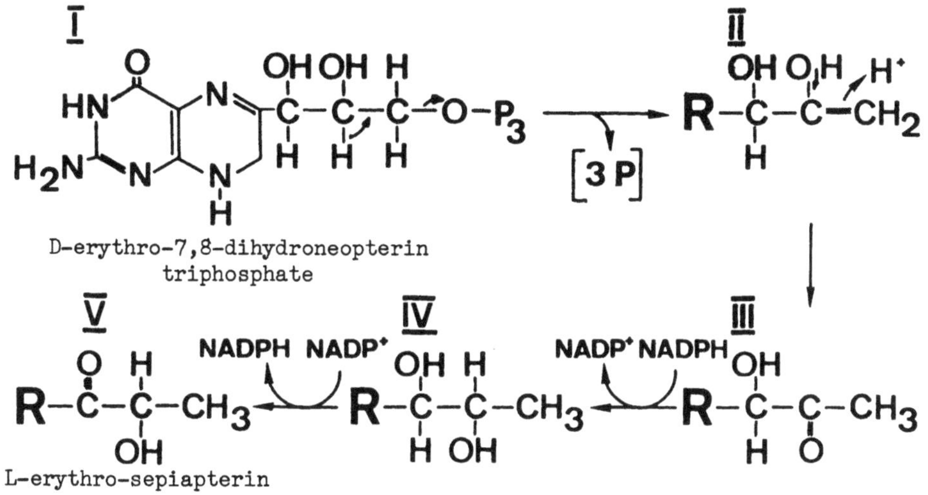

Fig. 1. Reaction scheme of Krivi and Brown. Enzymatic reaction sequence for the conversion of NH_2TP to sepiapterin proposed by Krivi and Brown. Compounds II-IV: Proposed intermediates.

Otsuka and coworkers (25) investigated BH_4 biosynthesis in *Ascaris lumbricoide suum*. According to their findings (see figure 2), GTP was converted to NH_2TP by GTP-cyclohydrolase I. NH_2TP was then converted to sepiapterin by the enzyme, sepiapterin synthetase. Their proposed synthetic pathway is shown in figure 2.

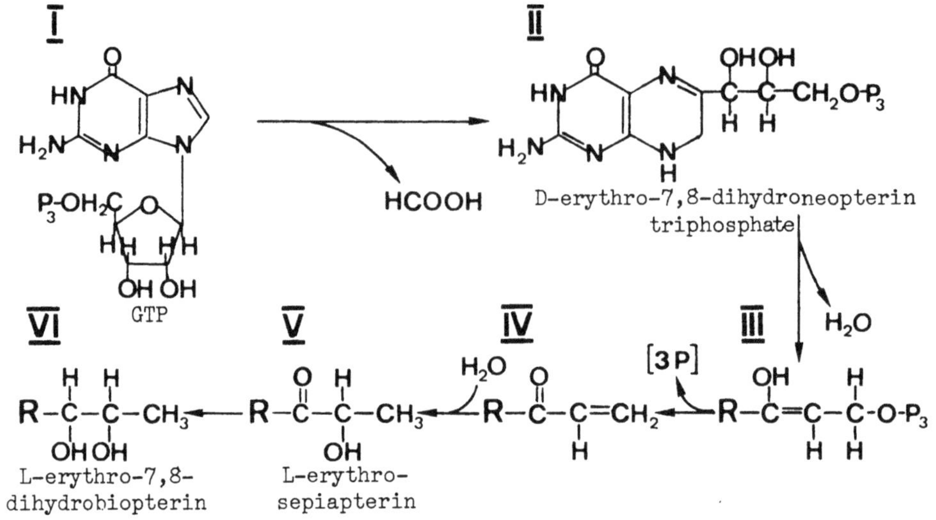

Fig. 2. The enzymatic conversions of GTP to BH_2 proposed by Otsuka and coworkers.
Compounds III and IV: Proposed intermediates.

Mammalian and avian systems

The details of the biosynthetic pathway of BH_4 in mammals are still under investigation. Gál and coworkers (26,27,28) investigated the BH_4 biosynthetic enzymes using rat brain extracts (see figure 3). They proposed that the initial conversion of GTP was catalyzed by one of two similar enzymes, GTP-cyclohydrolase A-I or Mg^{++}-dependent A-II, and that this reaction was rate-limiting in BH_4 biosynthesis. Both of these enzymes hydrolyzed GTP to 2-amino-6-(5-triphosphoribosyl)-amino-5-(or -6)-formamido-4-oxopyrimidine ($FPydP_3$). This compound was cyclized by D-erythro-7,8-dihydroneopterin triphosphate synthetase into quinonoid D-erythro-dihydroneopterin triphosphate ($q\text{-}NH_2TP$). $q\text{-}NH_2TP$ was subsequently transformed into quinonoid-L-erythro-dihydrobiopterin ($q\text{-}BH_2$) by the action of L-erythro-dihydrobiopterin synthetase. According to Gál, this reaction was catalyzed by a single enzyme that did not require pyridine nucleotides or Mg^{++} for catalytic activity. The biosynthesis of BH_4 as proposed by Gál is shown in figure 3. The details displayed in figure 3 are contrary to results in mammalian systems that were obtained by other laboratories.

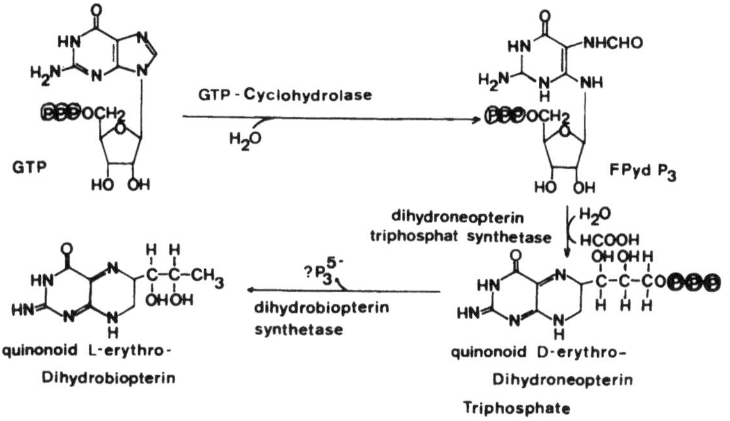

Fig. 3. Postulated biosynthesis of q-dihydrobiopterin in rat brain proposed by Gál and coworkers.

Fukushima and coworkers (29) have studied a NH_2TP synthetase (also called GTP-cyclohydrolase I) from chicken liver. This enzyme catalyzes the direct conversion of GTP to D-erythro-NH_2TP. In the process, the carbon atom from position 8 of GTP is released as formic acid. In other studies on rat liver GTP-cyclohydrolase I, Dhondt and coworkers (30) were able to fractionate 3 different forms of the enzyme. One form was heat stable and catalyzed the synthesis of NH_2TP in the presence of EDTA.

In human liver, our group has recently shown that in the presence of EDTA only one protein fraction synthesizes NH_2TP from

GTP (31). The partial purification was done by ammonium sulfate precipitation (30-50%) and column chromatography on DEAE Sephacel. This enzyme fraction is heat stable and shows no substrate inhibition.

Investigations in several laboratories on the enzymatic transformation of NH_2TP have provided generally consistent results. Tanaka and coworkers (32) obtained three protein fractions from chicken kidney, referred to as A1, A2, and B, that catalyze the conversion of NH_2TP to BH_2 in the presence of Mg^{++} and NADPH. Fraction A2, which was heat-resistant, catalyzed the conversion of NH_2TP in the presence of Mg^{++} to an unknown intermediate referred to as compound X. Fraction A1 catalyzed the subsequent conversion of compound X to sepiapterin in the presence of NADPH. Fraction B contained sepiapterin reductase (33), which catalyzed the conversion of sepiapterin to BH_2 in the presence of NADPH. Fraction A1 was demonstrated to be heat labile. These reactions are shown in figure 4.

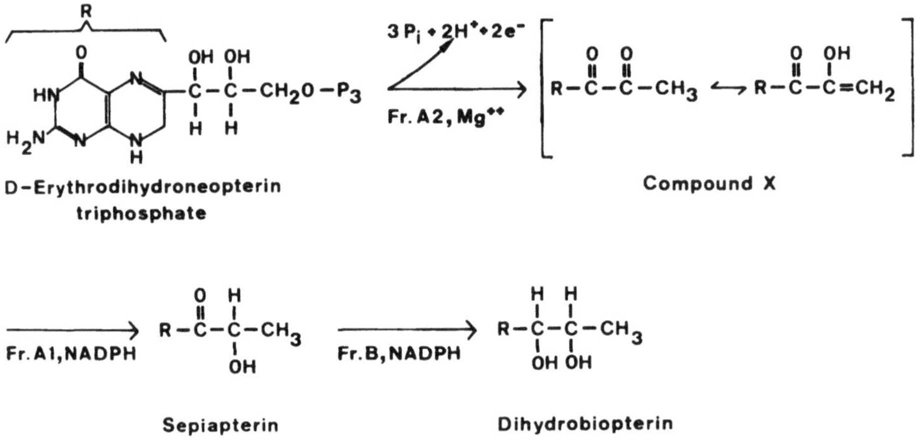

Fig. 4. Initial proposal by Tanaka and coworkers for the biosynthetic pathway of BH_2.

The pathways of Tanaka and coworkers in chicken kidney (32) and of Krivi and Brown (24) in *Drosophila melanogaster* are very similar. Both systems catalyze the first reaction from GTP to NH_2TP with only one enzyme fraction. The enzymes converting NH_2TP into sepiapterin have similar pH optima, K_M values and molecular weights. Both groups of investigators postulate a labile intermediate product and the enzymes in both systems require Mg^{++} ions as well as NADPH. But Krivi and Brown have not followed the assumption of an oxidation in the C-1' position followed by oxidation in the C-2' position. They suggested that the enzyme fraction A catalyzed the splitting of the phosphate groups via an elimination reaction that led to a labile intermediate referred to as compound III. This compound is thought to be 6-(1'-hydroxy, 2'-oxopropyl)-7,8-dihydropterin. Enzyme B is thought to reduce this intermediate to L-threo-BH_2 in the presence of NADPH, followed by an oxidation to sepiapterin.

Kapatos and coworkers described a cell-free system prepared from rat brain striatum that was capable of synthetizing BH_4 from GTP (34). They have provided evidence that sepiapterin is an intermediate in BH_4 synthesis.

Table 1 provides a summary of selected studies on BH_4 biosynthesis in a wide variety of systems.

Recent Observations on BH_4 Biosynthesis in Mammalian Systems

Experiments in this laboratory have dealt with BH_4 synthesis in liver and kidney extracts from man and rat (36,37). Following a procedure similar to that of Tanaka and coworkers, incubation of [U-^{14}C]NH_2TP with partially purified fractions of liver and kidney homogenates from man and rat yielded ^{14}C-labeled sepiapterin, BH_2, and compound X (indirectly measured),

Table 1

Comparison of different pathways of biopterin biosynthesis in mammalian and non-mammalian systems

System	Substrate	Enzyme	Products	Properties of enzyme and remarks
E.coli (18)	GTP	GTP-cyclohydrolase I	NH_2TP and formic acid	optimal pH 8-9; K_M for GTP=0.02 μM; M.W.=210,000; no cofactor req.
Comamonas sp. (19)	GTP	GTP-cyclohydrolase I	NH_2TP and formic acid	Mg^{++} req.; K_M for GTP=20 μM; M.W.=650,000
Lactobacillus plantarum (21)	GTP	NH_2TP synthetase	NH_2TP and formic acid	M.W.=200,000; stabilized by phosphate ions
Serratia indica (20)	GTP	GTP-cyclohydrolase D-I and D-II	NH_2TP	D-I: M.W.=200,000; K_M for GTP=32 μM at low GTP conc., 0.61 μM at high GTP conc. D-II: M.W.=200,000; K_M for GTP=0.44 μM at low GTP conc., 5.5 μM at high GTP conc.; inhibited by bi- and trivalent cations; K^+,

Table 1 cont (1):

				Na^+ and Li^+ are stimulating
Ascaris lumbricoides suum (25)	NH_2TP	sepiapterin synthetase	L-sepiapterin	
Drosophila melanogaster (23)	GTP	GTP-cyclohydrolase I = NH_2TP synthetase	NH_2TP and formic acid	optimal pH 8.7; K_M for GTP 22 µM; M.W.=345,000; no cofactors req.
	NH_2TP	enzyme A	compound III (see fig. 1)	req. Mg^{++}; M.W.=82,000; proposed structure is 6-(1'-hydroxy, 2'-oxopropyl)-7,8-dihydropterin
	compound III (see fig. 1)	enzyme B	sepiapterin	req. NADPH; M.W.=36,000
<u>Chicken</u>				
- liver (29)	GTP	GTP-cyclohydrolase I = NH_2TP synthetase	NH_2TP and formic acid	optimal pH 8.9; K_M for GTP 14 µM; M.W.=125,000; no cofactors req.

Table 1 cont (2):

- kidney (32)	NH_2TP	enzyme fraction A2	compound X (see fig. 8)	req. Mg^{++}; M.W.=77,000; proposed structure is 6-(1',2'-dioxopropyl)-7,8-dihydropterin
	compound X	enzyme fraction A1	L-sepiapterin	req. NADPH; M.W.30,000
	L-sepiapterin	sepiapterin reductase	L-dihydrobiopterin	req. NADPH
Syrian golden hamster (44)				
- kidney	NH_2TP	probably several	L-dihydrobiopterin	req. Mg^{++}, NADPH or NADH
Rat				
- brain (45)	NH_2TP	probably several	L-dihydrobiopterin	req. Mg^{++} and NADPH
- striatum (34)	GTP	probably several	L-sepiapterin	req. Mg^{++} and NADPH
- brain (26)	GTP	GTP-cyclohydrolase A-I or A-II	$FPydP_3$	A-I: optimal pH 8.0; K_M for GTP=1.2 μM; M.W.=135,000; no cofactors req. A-II: K_M for GTP=0.58 μM; M.W.=34,000; Mg^{++} req.
	$FPydP_3$	NH_2TP synthetase	NH_2TP	optimal pH 8.0; K_M for $FPydP_3$=1.4 μM; M.W.=9,080

Table 1 cont (3):

	NH_2TP	dihydrobiopterin synthetase	dihydrobiopterin	optimal pH 8.0; K_M for NH_2TP 17 μM; M.W.=124,000; no reqs for cofactors
Human				
- liver (31)	GTP	GTP-cyclohydrolase I	NH_2TP and formic acid	
- liver and kidney (37)	NH_2TP	enzyme fraction A2	compound X	req. Mg^{++}
	compound X	enzyme fraction A1	L-sepiapterin	req. NADPH
	L-sepiapterin	sepiapterin reductase	L-dihydrobiopterin	req. NADPH

depending on the particular assay conditions used. These findings are in support of the hypothesis that in rats as well as humans, BH_4 biosynthesis proceeds via sepiapterin.

Preparation of enzymes and substrates involved in sepiapterin and BH_2 synthesis in man and rat

Our procedure for studying sepiapterin synthesis is summarized in the following description: $[U-^{14}C]NH_2TP$ was used as substrate for the enzyme assay. $[U-^{14}C]NH_2TP$ was synthesized from $[U-^{14}C]GTP$ via E.coli GTP-cyclohydrolase I (36) and used without further purification. Between 5 and 10 g of liver or kidney tissue from human as well as from rat were used as a source of enzymes to study sepiapterin synthesis. The tissue was homogenized, centrifuged at 27'000 x g and the supernatant was subjected to ammonium sulfate precipitation (40-65%). The precipitate was redissolved, dialyzed, and the solution was applied on an Ultrogel AcA 34 column. As described previously, Tanaka and coworkers demonstrated that components A1 and A2 of fraction A from chicken kidney are not resolved by the gel filtration step, however, the A2 component of fraction A was heat stable. Since the A1 component was heat inactivated and was necessary for sepiapterin formation, this made it possible to study the presence of the intermediate, compound X. Thus, when studying the synthesis of compound X, we heated fraction A from the Ultrogel column to remove the A1 component. To investigate the formation of sepiapterin, fraction A was not heated, so that component A1 and A2 were both active. Although Tanaka and coworkers reported a complete separation of fraction A and B (sepiapterin reductase) on Ultrogel, we consistently were unable to separate completely fraction A and B. Nevertheless, a partial separation of fractions A and B was achieved. In the following experiments, Ultrogel fractions from liver and kidney extracts were incubated with $[U-^{14}C]$ NH_2TP and the formation of either compound X or sepiapterin and BH_2 was monitored as described below.

Assay for compound X

As mentioned previously, the labile intermediate, compound X, was formed from [U-^{14}C]NH_2TP by fraction A2 after A1 was destroyed by heat treatment (see figure 5).

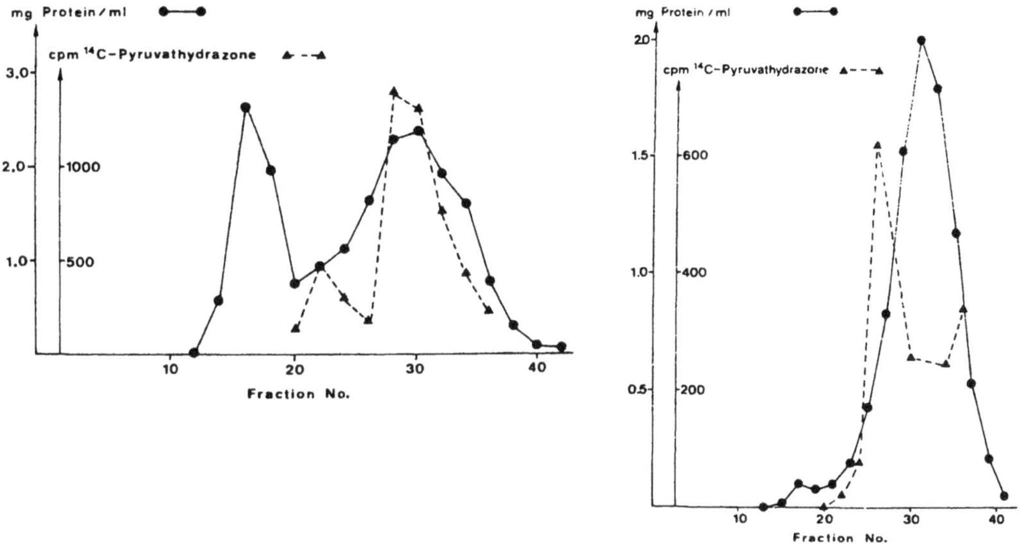

Fig. 5. Elution profile of Ultrogel AcA 34 enzyme preparation from rat liver (A) and human kidney (B). Enzyme A activity was determined by incubating the NH_2TP preparation with the Ultrogel AcA 34 fractions and measurement of the isolated [^{14}C]pyruvate hydrazone.

Following incubation of [U-^{14}C]NH_2TP with fraction A2, dinitrophenylhydrazine was added immediately to the reaction. If there is a diketo compound X, upon reaction with dinitrophenylhydrazine, a dinitrophenyl pyruvate hydrazone should be formed. This would provide indirect evidence for the existence of compound X. In fact, the pyruvate hydrazone could be demonstrated as described below. It was extracted with chloroform and separated by TLC in 1-butanol:acetic acid:water (20:1:3, v/v) on silica gel. The zone that comigrated with authentic pyruvate hydrazone was scraped off and eluted with ethanol:water (80:20, v/v) and the radioactivity of the eluates was quantitated.

To provide evidence that the hydrazone was indeed formed from pyruvic acid, the hydrazone was cleaved under reducing conditions so that pyruvate can be converted to alanine; the dinitrophenyl pyruvate hydrazone was hydrogenated catalytically with platinum oxide in ethanol at pH 7. Identification of radioactive alanine was done by TLC in 1-butanol:acetic acid:water (4:1:1, v/v) on silica gel, by a comparison with authentic material. The ability of rat liver and human kidney extracts to form the pyruvate hydrazone is shown in figure 5.

Assay for sepiapterin and BH_2 formation

Fractions from the Ultrogel AcA 34 column were eluted as described above, however, they were not heated in order to retain component A1 enzyme activity. Since we could not achieve complete separation of fractions A and B on Ultrogel, it was necessary to test the ability of these fractions to convert [U-^{14}C]NH_2TP into either sepiapterin, BH_2, or both (see figure 6). A portion of each fraction was reacted in the dark for 1 hour at 37°C with the following mixture in a final volume of 1 ml at pH 7.4: 1) 0.05 M Tris buffer (pH 7.4); 2) 20-200 nmol [U-^{14}C]NH_2TP prepared as previously described; 3) 2.5 mM NADPH; and 4) 0.5 mM $MgSO_4$. The reaction was terminated by applying the reaction products over a Ecteola-Sephadex G-25 column (16x320 mm) equilibrated with water. Standard pterins were eluted in the following order: BH_2, sepiapterin, biopterin, and pterin. The radioactivity from samples that co-eluted with standard BH_2 and sepiapterin was collected and pooled for further purification by HPLC (see figure 7). Radioactive

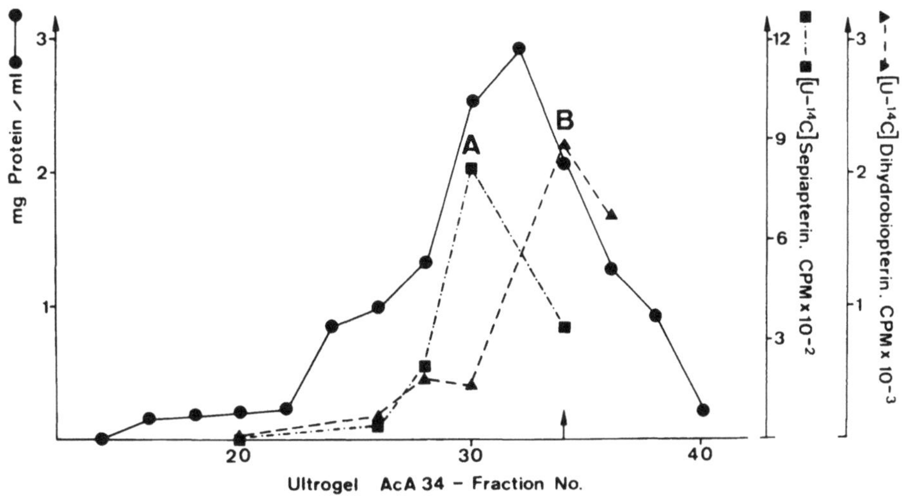

Fig. 6. Elution profile of 2 ml of a 40-65% saturated ammonium sulfate fraction of human kidney on an Ultrogel AcA 34 column (16x580 mm) in 0.05 mol/l potassium phosphate pH 6.8 . Flow rate 10 ml/h, 3 ml fractions.

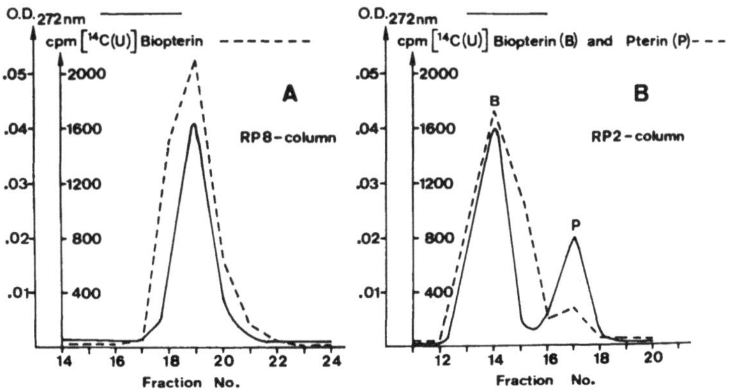

Fig. 7. HPLC purification of [U-^{14}C]B and [U-^{14}C] pterin synthesized by rat kidney extracts.

A: Second HPLC chromatography on Lichrosorb RP-8 of pooled fractions (B + pterin) from the first chromatography on RP-8.

B: Elution profile of the pooled fractions from A (18 - 21) on Lichrosorb RP-2.
B = Biopterin, P = Pterin.

fractions corresponding to the elution of BH_2 and sepiapterin were collected, concentrated and further analyzed by TLC after oxidation of BH_2 to biopterin by manganese dioxide at pH 2. The TLC system used was 1-butanol:acetic acid:water (20:3:7, v/v) on silica gel plates. After this third and final chromatographic purification, the radioactive BH_2 and sepiapterin were eluted from the scraped silica gel fractions with ethanol:3% ammonia (1:1, v/v) and radioactivity was quantitated. The results presented above are summarized in Table 2.

Table 2

Products of [U-^{14}C]GTP conversion by human kidney and rat liver *in vitro.*

Material	Ultrogel fraction	[U-^{14}C]GTP (dpm x 10^{-5})	Product	dpm/h / mg protein	% from GTP
1) Human kidney	A, 30[a]	7.3	[U-^{14}C]S[c]	367	0.05
	B, 34[a]	7.3	[U-^{14}C]BH_2	1200	0.16
2) Human kidney	A, 26[b]	4.4	[U-^{14}C]Pyr. hyd.[d] from compound X	2025	0.46
3) Rat liver	A, 28[b]	4.4		3720	0.85

[a]Fraction number refers to figure 6.
[b]Fraction numbers correspond to those from different experiments
[c]Sepiapterin
[d]Pyruvate hydrazone

Pterin measurements in human urine as an indicator of defects in BH_4 biosynthesis

Examination of the urinary pterin profile can often yield valuable information on the type of PKU patient and the proper method of treatment for the patient. Furthermore, clinical analyses can often substantiate results obtained from *in vitro* studies. In our PKU studies, we demonstrated relatively large amounts of 3'-hydroxysepiapterin (HS) in BH_2-deficient patients (3.9-9.6 mmol HS/mol creatinine) as compared with the low concentrations that were also detected in healthy controls as well as in patients with dihydropteridine reductase (DHPR) deficiency (40,41). It is assumed that the excreted HS is derived from HS triphosphate by cleavage of the phosphates. It is also possible that HS might be formed from NH_2TP or NH_2 by non-enzymatic oxidation, but this is not probable for two reasons. Firstly, BH_2 can also be oxidized into sepiapterin by a non-enzymatic oxidation, although only trace amounts of sepiapterin have been detected in the urine of DHPR-deficient patients. Secondly, after loading with NH_2TP (2.5 mg/kg), a high voltage electrophoresis chromatogram of fresh urine from an adult human control showed only traces of HS, but large amounts of NH_2 and neopterin were measured. However, after storage of this urine for 9 months at $-20^{o}C$, an HPLC chromatogram displayed remarkable concentrations of HS, thus, indicating that non-enzymatic oxidation had occured. Therefore it seems likely that HS triphosphate is on the metabolic pathway and an intermediate in the BH_4 biosynthesis. Thus, the most probable reaction sequence based on the above observations by Tanaka and other researchers is shown in figure 8.

Discussion

There are still many unanswered questions and several controversies regarding BH_4 biosynthesis. Following the investigations of Fukushima *et al.* in chicken liver (29), Dhondt (30) and Blau and Niederwieser (31) in human liver, there exists only one enzyme, GTP-cyclohydrolase I, synthetising NH_2TP from GTP. This is in agreement with the results of other authors investigating non-mammalian systems. On the other hand, Gál and coworkers claimed that in rat brain, there existed two distinct enzyme fractions converting GTP, first to a pyrimidine derivative ($FPydP_3$) (cyclohydrolase I and II) and a second enzyme fraction converting the pyrimidine to dihydroneopterin triphosphate (di-

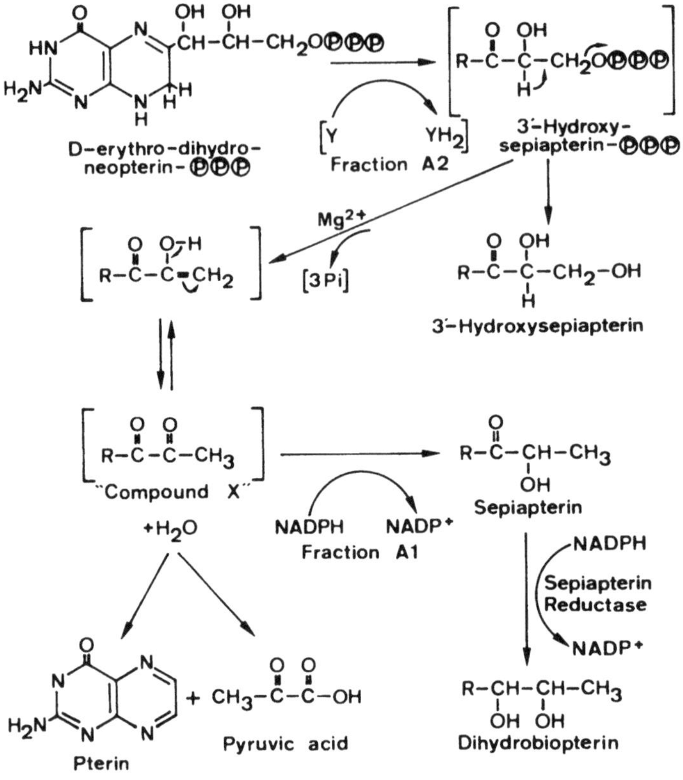

Fig. 8. Currently proposed biosynthetic pathway of dihydrobiopterin. This pathway is compiled from the work of Tanaka and coworkers and Niederwieser and coworkers.

hydroneopterin triphosphate synthetase) (26).

The existence of the intermediate, compound X, in the formation of sepiapterin from NH_2TP has been suggested by Tanaka and co-workers (32). While this has not been proven directly, there is growing evidence supporting the existence of this intermediate. The formation of the pyruvate hydrazone from the degradation of compound X indicates that the diketo side chain proposed for compound X could exist. The demonstration by Niederwieser *et al.*

(40) that large amounts of HS can be detected in the urine of a BH_2-deficient atypical PKU patient also supports the hypothesis of compound X. This observation further suggests that HS is an intermediate in the synthesis of BH_4.

Several groups have demonstrated that sepiapterin is converted to BH_2 by sepiapterin reductase, which indicates that sepiapterin may be an intermediate in BH_4 biosynthesis. However, the BH_2 formed appears to be in the non-quinonoid form and therefore, would not be reduced to BH_4 by quinonoid dihydropterin reductase. It has been shown *in vitro* that the conversion of BH_2 to BH_4 can be accomplished by dihydrofolate reductase (43), but whether this is a major reaction *in vivo* remains unclear.

It is noteworthy that neopterin cannot be detected in rat tissues (42), yet it exists in humans. Since the biosynthetic reactions appear to be similar in the rat and human but the pterin metabolic profile is different, this may indicate that GTP-cyclohydrolase is rate-limiting only in the rat and not in humans. The presence of neopterin in humans indicate that the conversion of NH_2TP to sepiapterin might be an important point of regulation in BH_4 synthesis in man.

Acknowledgements

We wish to thank Drs T. Shiota (University of Alabama, Birmingham, AL, USA) and E.M. Gál (Iowa City, IO, USA) for valuable discussions, and to Mr. W. Leimbacher and Miss A. Matasović for their technical assistance. We are especially grateful to Dr. R.A. Levine (NIH, Bethesda, MD, USA) for valuable discussions and his most helpful assistance in the preparation of this manuscript. This work was supported by the Swiss National Science Foundation project no. 3815-0.79.

References

1. Danks, D.M.: Pteridines and phenylketonuria. Report of a workshop. Introductory comments. J. Inher. Metab. Dis. 1, 47-48 (1978).

2. Kaufman, S., Holzman, N.A., Milstien, S., Buther, I.J. and Krumholz, A.: Phenylketonuria due to a deficiency of dihydropteridine reductase. New Engl. J. Med. 293, 785 (1975).

3. Niederwieser, A., Curtius, H.-Ch., Bettoni, O., Bieri, J., Schircks, B., Viscontini, M. and Schaub, J.: Atypical phenylketonuria caused by 7,8-dihydrobiopterin synthetase deficiency. Lancet 131-133 (1979).

4. Curtius, H.-Ch., Niederwieser, A., Viscontini, M., Otten, A., Schaub, J., Scheibenreiter, S., Schmidt, H.: Atypical phenylketonuria due to tetrahydrobiopterin deficiency.Diagnosis and treatment with tetrahydrobiopterin, dihydrobiopterin and sepiapterin. Clin. Chim. Acta 93, 251-262 (1979).

5. Niederwieser, A., Staudenmann, W., Wang, M., Curtius, H.-Ch., Atares, M., Cardesa-Garcia, J.: Hyperphenylalaninemia with neopterin deficiency. A new enzyme defect presumably of GTP-cyclohydrolase. Eur. J. Pediat. 138, 97 (1982).

6. Rey, R., Harpey, J.P., Leeming, R.J., Aicardi, J., Rey, F.: Les hyperphénylalaninémies avec activité normale de la phénylalanine hydroxylase. Arch. franç. Pédiat. 34 (suppl.) 109 (1977).

7. Bartholomé, K.: A new molecular defect in phenylketonuria. Lancet II, 1580 (1974).

8. Smith, I., Clayton, B.E., Wolff, O.H.: New variant of phenylketonuria with progressive neurological illness unresponsive to phenylalanine restriction. Lancet 1108-1111 (1975).

9. Kaufman, S.: Properties of the pterin-dependent aromatic amino acid hydroxylases. In: Aromatic Amino Acids in the Brain. CIBA Foundation, Elsevier, New York, pp 81-115 (1980).

10. Lovenberg, W. and Victor, S.J.: Regulation of tryptophan and tyrosine hydroxylase. Life Sci. 14, 2337-2353 (1974).

11. Niederwieser, A., Curtius, H.-Ch., Gitzelmann, R., Otten, A., Baerlocher, K., Blehova, B., Berlow, S., Gröbe, H., Rey, F., Schaub, J., Scheibenreiter, S., Schmidt, H. and Viscontini, M.: Excretion of pterins in phenylketonuria

and phenylketonuria variants. Helv. paediat. Acta 35, 335-342 (1980).

12. Niederwieser, A., Curtius, H.-Ch., Wang, M., Leupold, D.: Atypical phenylketonuria with defective biopterin metabolism. Monotherapy with tetrahydrobiopterin or sepiapterin, screening and study of biosynthesis in man. Eur. J. Pediat. 138, 110-112 (1982).

13. Curtius, H.-Ch., Müldner, H. and Niederwieser, A.: Tetrahydrobiopterin effective in endogenous depression and Parkinson's disease. (Submitted for publication).

14. Curtius, H.-Ch., Niederwieser, A. and Müldner, H.: Tetrahydrobiopterin effective in Parkinson syndrome. Proc. of the VIIth Int. Symp. on Parkinson's disease, in press.

15. Nagatsu, T., Yamaguchi, T., Rahman, K., Trocewicz, J., Oka, K., Nagatsu, I., Narabayashi, H., Kondo, T. and Iizuka, R.: Catecholamine-related enzymes and the biopterin cofactor in Parkinson's disease. Proc. of the VIIth Int. Symp. on Parkinson's disease, in press.

16. LeWitt, P., Miller, L., Insel, Th., Calne, D., Lovenberg, W. and Chase, Th.: Tyrosine hydroxylase cofactor in Parkinsonism. Proc. of the VIIth Int. Symp. on Parkinson's disease, in press.

17. Burg, A.W. and Brown, G.M.: The biosynthesis of folic acid. VIII. Purification and properties of the enzyme that catalyzes the production of formate from carbon atom 8 of guanosine triphosphate. J. Biol. Chem. 243, 2349-2358 (1968).

18. Yim, J.J. and Brown, G.M.: Characteristics of guanosine triphosphate cyclohydrolase I purified from E.coli. J. Biol. Chem. 251, 5087-5094 (1976).

19. Cone, J. and Guroff, G.: Partial purification and properties of guanosine triphosphate cyclohydrolase, the first enzyme in pteridine biosynthesis, from *Comamonas sp.* (ATCC 112299a) J. Biol. Chem. 246, 979-985 (1971).

20. Kohashi, M., Itadani, T. and Iwai, K.: Purification and characterization of guanosine triphosphate cyclohydrolase from *Serratia indica*. Agric. Biol. Chem. 44 (2), 271-278 (1980).

21. Jackson, R.J. and Shiota, T.: The nature of the multiple forms of D-erythro-dihydroneopterin triphosphate synthetase. Biochim. Biophys. Acta 403, 232-244 (1975).

22. Suzuku, Y., Yasui, T. and Abe, S.: Occurence of GTP cyclohydrolase I in *Bacillum stearothermophilus*. J. Biochem. 86, 1679-1685 (1979).

23. Fan, C.L. and Brown, G.M.: Partial purification and properties of guanosine triphosphate cyclohydrolase from *Drosophila melanogaster*. Biochem. Gen. 14 (3/4), 259-270 (1976).

24. Krivi, G.G. and Brown, G.M.: Purification and properties of the enzymes from *Drosophila melanogaster* that catalyze the synthesis of sepiapterin from dihydroneopterin triphosphate. Biochem. Gen. 17, 371-390 (1979).

25. Otsuka, H., Suguira, K. and Goto, M.: Biosynthesis of biopterin in *Ascaris lumbricoides suum*. Biochim. Biophys. Acta 629, 69-76 (1980).

26. Gál, E.M., Nelson, J.M. and Sherman, A.D.: Biopterin. III. Purification and characterization of enzymes involved in the cerebral synthesis of 7,8-dihydrobiopterin. Neurochem. Res. 3, 69-88 (1978).

27. Gál, E.M., Henn, F.A. and Sherman, A.: Biopterin IV. Regional and subcellular aspects of L-erythro-7,8-dihydrobiopterin synthesis in brain. Neurochem. Res. 3, 493-499 (1978).

28. Gál, E.M., Dawson, M.R., Dudley, D.T. and Sherman, A.D.: Biopterin: VI. Purification and primary amino acid sequence of mammalian D-erythro-7,8-dihydroneopterin triphosphate synthetase. Neurochem. Res. 4 (5), 605-625 (1979).

29. Fukushima, K., Richter, W.E. Jr., Shiota, T.: Partial purification of 6-(D-erythro-1',2',3'-trihydroxypropyl)-7,8-dihydropterin triphosphate synthetase from chicken liver. J. Biol. Chem. 252, 5750-5755 (1977).

30. Dhondt, J.L.: personal communication.

31. Blau, N., and Niederwieser, A.: GTP-cyclohydrolase I assay in human and rat liver using HPLC of neopterin phosphates and guanine nucleotides. Anal. Biochem., submitted for publication.

32. Tanaka, K., Akino, M., Hagi, Y., Doi, M. and Shiota, T.: The enzymatic synthesis of sepiapterin by chicken kidney preparations. J. Biol. Chem. 256, 2963-2972 (1981).

33. Matsubara, M., Katoh, S., Akino, M. and Kaufman, S.: Sepiapterin reductase. Biochim. Biophys. Acta 122, 202-212 (1966).

34. Kapatos, G., Katoh, S. and Kaufman, S.: Sepiapterin (6-lactyl-7,8-dihydropterin) is an intermediate in biopterin biosynthesis in a cell-free preparation from rat striatum. In: Function and Regulation of Monoamine Enzymes (Usdin, E., Weiner, N., Youdim, M.B.H., eds.) MacMillan Publ. Ltd. London, pp 263-270 (1981).

35. Tanaka, K., Akino, M., Hagi, Y. and Shiota, T.: Biosynthesis of biopterin by chicken enzyme preparations. In: Chemistry and Biology of Pteridines. (Kisliuk, R.L. Brown, G.M., eds.) Elsevier/North-Holland, pp 147-152 (1979).

36. Curtius, H.-Ch., Häusermann, M., Niederwieser, A., Endres, W. and Viscontini, M.: New aspects in biopterin biosyntheses. In: Function and Regulation of Monoamine Enzymes. (Usdin, E., Weiner, N., Youdim, M.B.H., eds) MacMillan Publ. Ltd. London, pp 251-262 (1981).

37. Häusermann, M., Ghisla, S., Niederwieser, A. and Curtius, H.-Ch.: New aspects in biopterin biosynthesis in man. FEBS Lett. 131 (2), 275-278 (1981).

38. Viscontini, M. and Furuta, Y.: Synthese von D-Neopterin-3'-phosphat und D-Neopterin-2',3'-cyclophosphat. Betrachtungen über die Biopterin-Biogenese. Helv. Chim. Acta 56 1819-1821 (1973).

39. Antoulas, S. Viscontini, M.: Synthèse de l'ester tripolyphosphorique de la tétrahydro-5,6,7,8-D-néoptérine. Essai d'interprétation de la biosynthèse de la L-sépiaptérine. Helv. Chim. Acta 64, 1134 (1981).

40. Niederwieser, A., Matasović, A., Curtius, H.-Ch., Endres, W. and Schaub. J.: 3'-Hydroxysepiapterin in patients with dihydrobiopterin deficiency. FEBS Lett. 118, 299-302 (1980).

41. Curtius, H.-Ch., Häusermann, M., Niederwieser, A., Endres, W. and Viscontini, M.: New aspects in biopterin biosyntheses. In: Function and Regulation of Monoamine Enzymes (Usdin, E., Weiner, N., Youdim, M.B.H., eds.) Macmillan Publ. Ltd. London, pp 251-262 (1981).

42. Fukushima T., Nixon, J.C.: Analysis of reduced forms of biopterin in biological tissues and fluids. Anal. Biochem. 102, 176-188 (1980).

43. Abelson, H.T., Spector, R., Gorka, C. and Fosburg, M.: Kinetics of tetrahydrobiopterin synthesis by rabbit brain dihydrofolate reductase. Biochem. J. 171, 267-268 (1978).

44. Eto, I., Fukushima, K. and Shiota, T.: Enzymatic synthesis of biopterin from D-erythro-dihydroneopterin triphosphate by extracts of kidneys from Syrian golden hamsters. J. Biol. Chem. 251, 6505-6512 (1976).

45. Lee, C.-L., Fukushima, T. and Nixon, J.C.: Biosynthesis of biopterin in rat brain. In: Chemistry and Biology of Pteridines. (Kisluik, R.L., Brown, G.M., eds.) Elsevier/North-Holland, pp 125-128 (1979).

BALANCING OF MAMMALIAN BIOPTERIN POOLS

Heinz Rembold
Max Planck Institute for Biochemistry,
D-8033 Martinsried, Germany

Introduction

Unconjugated pterins of biopterin type (6-hydroxyalkyl-pterins) are obviously ubiquitous in nature (1). They are synthesized *de novo* also by the mammal, as has been proved *in vivo* with rats (2) which received a biopterin-free diet for several generations and which did not differ from the control in growth and biopterin excretion, and *in vitro* with mammalian cell cultures (3,4) which use GTP as a precursor. Biopterin is a growth factor for the trypanosomid flagellate, *Crithidia fasciculata*. Half maximum growth in the microbiological assay (5) is exhibited at a biopterin concentration of 2.5 x 10^{-8} mg/ml (6). A radioimmuno assay with a low percentage cross reaction of pterins with L-*erythro* biopterin antiserum has recently been described (7).

The biological function of tetrahydrobiopterin (BH_4) in the mammal as cofactor in several mixed function oxygenases has been shown for phenylalanine hydroxylase (8) and for the rate-limiting enzymes of catecholamine biosynthesis, i.e. tyrosine and tryptophan hydroxylase (9). This suggests that control of the coenzyme site might have a long-term regulatory effect on functional neurotransmitter pools (10-12). The amount of the enzyme proteins may be increased as a consequence of the induction of these enzymes without changes in the Km values against the pterin cofactor. Therefore, the biosynthesis of the monoamine-synthesizing oxygenases, their substrates and

Biochemical and Clinical Aspects of Pteridines · Volume 1
Cancer · Immunology · Metabolic Diseases

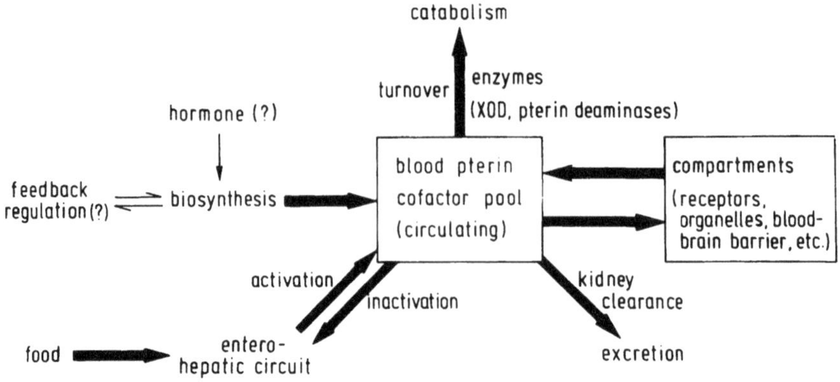

Fig. 1. Control scheme for pterin cofactor balancing through synthesis, degradation, compartmentation and excretion in the mammal.

end products, and that of the reduced pterin cofactor may be regulated with each other.

By what other means can the mammalian organism balance its pterin-cofactor levels? One answer can be expected from clinical manifestations of a disturbed biopterin biosynthesis or its metabolites (13). For a normal regulation of pterin cofactor pools, however, we can expect quite different pathways, as exemplified in the scheme of figure 1.

The mammal as an open biochemical system controls its reduced pterin cofactors at a constant level which differs from species to species. There are also remarkable differences within the organism from organ to organ, as exemplified in table 1 for the rat. The highest biopterin concentration is found in the liver (4.6 µmol/kg). If compared with plasma concentration (0.11 µmol/kg), the liver value demonstrates a 40fold uphill concentration. How such differences are maintained at a constant concentration is unknown. There is also unknown, how much of the pterin cofactor level in a certain compartment is maintained by cellular *de novo* synthesis or by an uphill transport through membrane barriers.

Tab. 1. Biopterin concentration (μmol/kg wet weight of different rat organs and blood according to *Crithidia* growth activity. SD = standard deviation, n = number of indipendent isolations (14).

Organ	$c_{biopterin}$	SD	n
liver	4.60	0.80	6
spleen	2.29	0.57	4
kidney (total)	1.03	0.43	21
(cortex)	1.20	0.35	10
(outer medulla)	1.10	0.33	10
brain	0.25	0.08	11
heart	0.13	0.02	4
erythrocytes	0.55		2
blood	0.25	0.04	4
plasma	0.11	0.025	7

These compartmentations must have a physiological significance which becomes more evident if a pulse of labelled tetrahydrobiopterin is injected (Fig. 2). One can immediately realize from the regional distribution of the label, that BH_4 does not pass the blood brain barrier in a substantial amount. Most of the radioactivity is concentrated in the liver which is in accordance with the high biopterin concentration of this organ (Tab. 1). From the rate of incorporation can be calculated, that the labelled BH_4 is resorbed by the liver against a concentration gradient. An appreciable amount of radioactivity is also located in the gastro-intestinal tract and in the kidney.

From the possible regulatory circuits as summarized in figure 1, only the biosynthetic pathway of BH_4 and its pathological manifestations in atypical PKU has attracted also the clinician. As this aspect is covered by other contributions to the present workshop, it will not be discussed in the following. It may just be mentioned that nothing is known at present

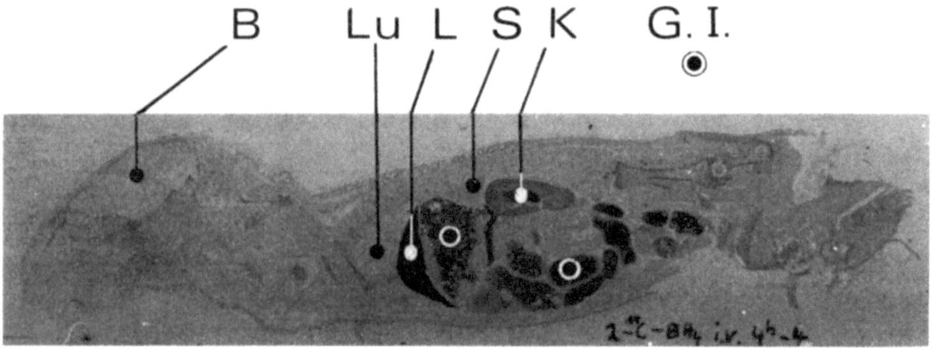

Fig. 2. Autoradiography after injection of 5 µCi 2-^{14}C-BH_4 into the tail vein of a male rat (150 g). The animal was killed by freezing after ether anaesthesia 4 h after injection and cut into 40 µm sections in a cryostate. B = brain, L = liver, S = spleen, Lu = lung, G.I. = gastro-intestinal tract, K = kidney (for more details see also loc.cit. 14).

about a feedback control of BH_4 biosynthesis and whether any hormonal control system is involved.

Compartmentation of pterin pools

As was demonstrated already by the data of table 1, biopterin and possibly other 6-hydroxyalkylpterins as well are contained in different concentrations in the higher organisms. In organs again, a functional compartmentation appears (Tab. 2). *Crithidia*-active material is concentrated in such parts of the rat brain, where also catecholamines and serotonin are located as neurotransmitters. Biopterin concentration of the cortex is comparatively low and the hypothalamus is the region with the highest concentration.

Such an organ specific concentration must be based on an interaction with receptor proteins, cell organelles, or membrane barriers like in the case of the blood brain barrier. By application of density gradient centrifugation in sucrose, *Crithidia*-active material could be separated from rat brain synaptosomes into the myelin containing, the synaptosomal and the mitochondrial fractions (14). An analysis of subcellular fractions demonstrated that most of *Crithidia* activity is contained in the cytosol. Distinct activity, however, was also found in the fractions of cell nuclei and also in the mitochondria.

Tab. 2. Regional distribution of biopterin (*Crithidia* activity) in the rat brain. Average values from 11 animals in µmol/kg wet weight (14). SD = standard deviation.

Anatomical region	$c_{biopterin}$	SD
brain (total)	0.25	0.08
cerebellum	0.18	0.07
medulla oblongata and pons	0.18	0.09
hypothalamus	0.49	0.11
midbrain	0.25	0.09
hippocampus	0.13	0.07
corpus striatum	0.22	0.02
cortex	0.07	0.02

The biological importance of BH_4, besides its cofactor function in oxygenases, is still not clear. There are some first data, however, which speak in favour of additional functions of pterin cofactors in intermediary metabolism and which fit in their above mentioned cellular and subcellular compartmentation. Rembold and Buff (11,12) suggest that reduced pterins have a mediator function in cellular electron transfer by linking together cytosolic reduced pyridine nucleotide pools with the respiratory chain of mitochondria at the cytochrome a/a_3 site. Cytochrome c would act as an electron shuttle between the outer and inner mitochondrial membranes in this system which would become active at extraordinarily high cytosolic NADPH/NADH-levels. This could be the case in regenerating, embryonic, or neoplastic tissues. There are some data which support this hypothesis (17,18) which becomes of special interest in connection with an increased BH_4 level in regenerating tissues (19,20) and in tumor-bearing organisms (21). Ebels and coworkers have studied the possible significance of pteridines in pineals and other neurosecretory tissues concerning the biosynthesis of melatonin (22), antigonadotropin (23) and other neurosecretions (24). It will need more detailed investigations, however, before these suggested pterin functions are clarified.

Binding capacity of human blood serum was found very high for tetrahydropterins (25). It is in the range of 5×10^{-10} to 10^{-3} M without saturation. Only at concentrations of less than 0.8×10^{-8} M high affinity binding seems to be possible which is about the physiological biopterin concentration in human blood (2×10^{-8} M). Specificity for tetrahydropterin binding sites was demonstrated by the authors in the α_2-macroglobulin fraction. A pteridine-binding α_1-acid glycoprotein has been isolated from blood of patients with neoplastic diseases (26,27). An uphill biopterin transport under energy consumption has been measured in *Crithidia fasciculata* (28). Whether here again high affinity receptor proteins are involved, has not been studied. Inhibition of biopterin incorporation by colchicine or lumicolchicine, which do not penetrate the cell membrane, suggests however a steric hindrance of the biopterin carrier system (29).

Resorption and excretion of pterins

Only little is known about biopterin resorption from the gastro-intestinal tract. A small amount of it may pass unaltered across the bowel and be excreted again, another one is probably reduced during transport (30,31). BH_4 does not pass the blood brain barrier readily (14,32); there seems to exist a gastrohepatic circuit, however (14). A selective resorption of 6-biopterin (2) can be interpreted as an energy-dependent transport by the rat intestine which has been demonstrated for *Crithidia* already (28,29,33).

The fundamental importance of a renal balancing of pterins has been studied by Häberle et al. (34,35) by a clearance and micropuncture study in the rat. In rat kidney, the rate of urinary biopterin and BH_4 excretion was found to be at least three times higher than the rate at which it is filtered. In microperfusion experiments in isolated proximal tubular segments *in vivo et situ*, no appreciable reabsorption was found. From the fact, that more biopterin leaves the kidney with urine plus venous blood the authors conclude, that reduced biopterin is synthesized *de novo* in this organ. It is evident from their findings, that most of the circulating blood pterins are excreted with the urine and the blood biopterin level is possibly maintained by synthesis in the kidney already (Tab. 3).

Tab. 3. Biopterin concentrations (*Crithidia* activity) in plasma of arterial and venous renal blood and in the urine of the rat (36). SD = standard deviation, n = number of experiments.

	concentration	SD	n
renal artery	(ng/ml) 24.9	6.6	7
renal vein	(ng/ml) 27.0	5.9	7
urine	(µg/ml) 2.04	1.06	8

The kidney seems to play a central role in the control of pterin cofactor balancing. Alterations in patients with severe renal insufficiency (37) support this suggestion. The microperfusion studies also suggest, that the pterin cofactor pool is primarily regulated through control of pterin cofactor biosynthesis.

Catabolism of pterin cofactors

For a complete balancing of the pterin cofactors, their turnover from the dihydro- and tetrahydro-forms could be linked to a catabolic pathway which would end up in an excretion or deposition of xanthopterin, isoxanthopterin and leucopterin as is the case in many insects and amphibia (1). The presence of xanthopterin in human urine has long been known. In a series of papers Rembold and coworkers (38-42) have described the degradation of 6-hydroxyalkyl-tetrahydropterins in the rat *in vivo et in vitro*. Their results are summarized in the schemes of figure 3 with tetrahydroneopterin (I) as an example for the class of enzyme cofactors. Tetrahydrobiopterin is degraded similarly after intraperitoneal injection into the rat (42) or by incubation with rat liver homogenates (41).

According to figure 3, the trihydroxypropyl side chain is oxidatively cleaved in a non-enzymatic reaction to form glyceraldehyde. The remaining 7,8-dihydropterin (VIa) can be oxidized by xanthine oxidase or it can serve as a substrate for the highly specific enzyme pterin deaminase. In the last case, which is exemplified in the lower part of figure 3, either 7,8-dihydrolumazine (XIa) or lumazine (XII) is formed, depending on the state of oxidation of the substrate for the deaminase. The products in both cases serve as further substrates for xanthine oxidase. End products of the catabolic pathway are either neopterin (IV) and its 7,8-dihydro form, dihydroxanthopterin (VIII), xanthopterin (IX), isoxanthopterin (XV) and leucopterin (XVI), or in the lumazine series 7,8-dihydro-6-hydroxylumazine (XIII), 6-hydroxylumazine (XIV), 7-hydroxylumazine (XVII) and 7,8-dihydroxylumazine (XVIII). These end products of pterin cofactor metabolism are excreted with the urine. Some of them are useful indicators of rapid or malignant cell proliferation. Excretion of 6-hydroxylumazine (XIII) by mice bearing the Ehrlich ascites tumor (45), of neopterin (IV) by patients with malignant tumors (46), or of 7,8-

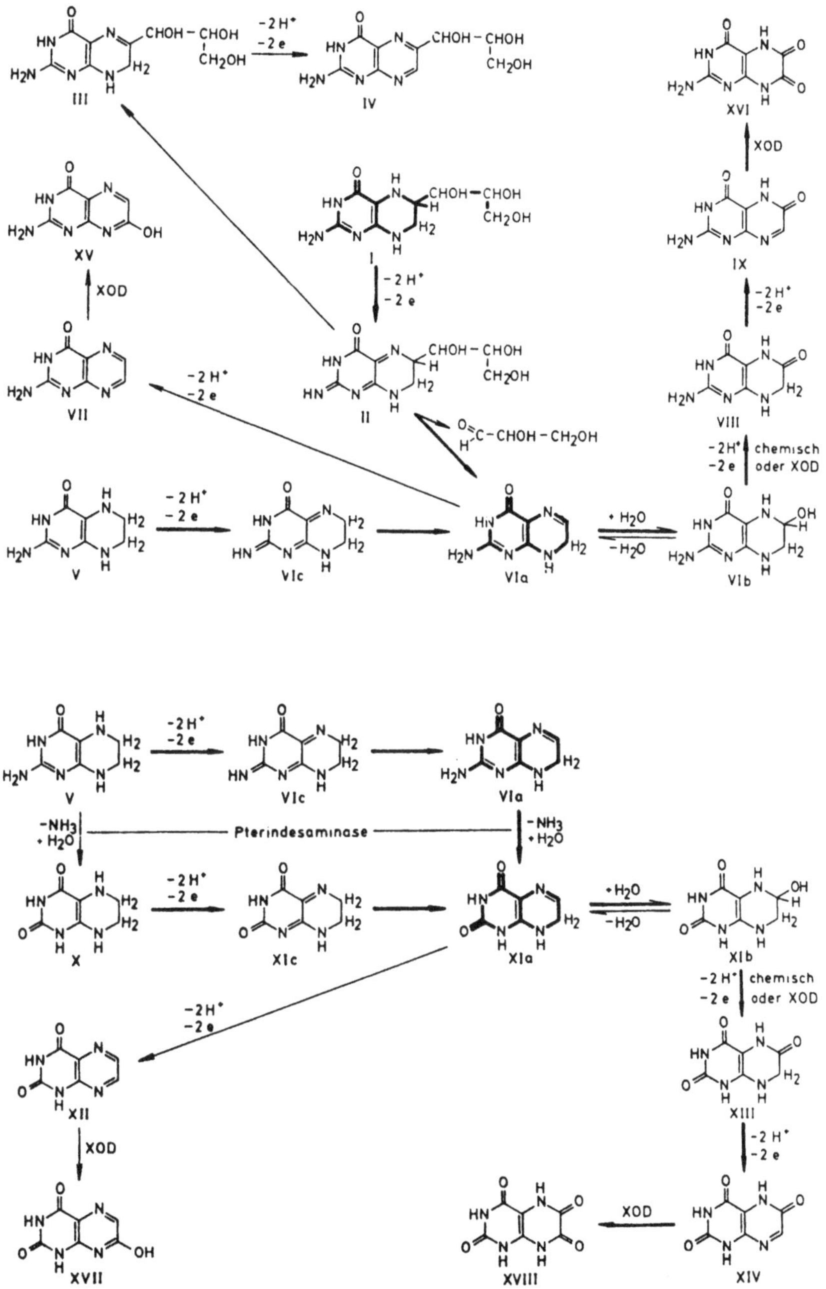

Fig. 3. Scheme for the catabolic pathway of pterin cofactors (from loc.cit. 43).

dihydroxanthopterin (VIII) (48) can now be used as a marker.

Conclusion

The scheme of figure 1 for the balancing of pterin cofactors in the mammalian organism is far from a complete understanding. There are first data, however, which support the idea of a dominant role of the biosynthetic pathway. The total pool size of biopterin is fairly low - in the range of µmol/kg - and it seems economic if this level is primarily controlled by compartmentation of the binding proteins. The physiological significance of the high-affinity biopterin binding protein, which was localized in the α_2-macroglobulin fraction of human blood serum, is not yet clear. It is remarkable, however, that its K_m value is in the range of blood biopterin concentration. In the circulating blood, the constant pterin cofactor level is controlled by a high filtration rate in the kidney with a concomitant release of new cofactor into the blood stream. If this view is correct, the kidney would have a dominant role in the balancing of mammalian biopterin pools. Supply of biopterin through the food seems to be of minor or no importance. An activation to the reduced cofactor has not been established till now, whereas administration of reduced cofactor, either oral or by injection (47) may substitute in anabolic defects. It is not yet clear, whether the catabolism of pterin cofactor plays a substantial role under normal metabolic conditions. In a pathological situation, however, like malignancy, PKU, viral diseases, the catabolic pathway, including the activation of pterin deaminases, may play an important role as an indicator of metabolic disorders for the clinician.

References

1. Rembold, H., Gyure, W.: Angew. Chem. 84, 1088-1099 (1972); Angew. Chem., Int. Ed. 11, 1061-1072 (1972).
2. Pabst, W., Rembold, H.: Z. physiol. Chem. 344, 107-112 (1966).
3. Fukushima, T., Shiota, T.: J. Biol. Chem. 249, 4445-4451 (1974).
4. Buff, K., Dairman, W.: Mol. Pharmacol. 11, 87-93 (1975).
5. Baker, H., Frank, O., Shapiro, A., Hutner, S.M.: Methods Enzymol. 66, 490-500 (1980).
6. Rembold, H., Metzger, H.: Z. physiol. Chem. 329, 291-292 (1962).
7. Nagatsu, T., Yamaguchi, T., Kato, T., Sugimoto, T., Matsuura, S., Akino, M., Tsushima, S., Nakazawa, N., Ogawa, H.: Analytical Biochemistry 110, 182-189 (1981).
8. Kaufman, S.: Proc. Natl. Acad. Sci. USA 50, 1085-1093 (1963).
9. Kaufman, S., Fisher, D.B.: In "Molecular Mechanisms in Oxygen Activation" (Hayaishi, D., ed.), pp. 285-369, Academic Press, New York 1974.
10. Levitt, M., Spector, S., Sjoerdsma, A., Udenfriend, S.: J. Pharm. Exp. Ther. 148, 1-8 (1965).
11. Grahame-Smith, D.G.: Biochem. J. 105, 351-360 (1967).
12. Friedman, P.A., Kappelman, A.H., Kaufman, S.: J. Biol. Chem. 247, 4165-4173 (1972).
13. Häusermann, M., Ghisla, S., Niederwieser, A., Curtius, M.-Ch.: FEBS Lett. 131, 275-278 (1981).
14. Hennings, G., Rembold, H.: Internat. J. Vit. Nutr. Res. 52, 36-43 (1982).
15. Rembold, H., Buff, K.: Eur. J. Biochem. 28, 579-585 (1972).
16. Rembold, H., Buff, K.: Eur. J. Biochem. 28, 586-591 (1972).
17. Taylor, D., Hochstein, P.: Biochem. Biophys. Res. Commun. 67, 156-162 (1975).
18. Harano, T.: Insect Biochem. 2, 385-399 (1972).
19. Kokolis, N.: Int. J. Biochem. 6, 475-478 (1975).
20. Kokolis, N.: Comp. Biochem. Physiol. 54B, 285-287 (1976).
21. Kokolis, N., Ziegler, I.: Cancer Biochem. Biophys. 2, 70-85 (1977).

22. Cremer-Bartels, G., Ebels, I.: Proc. Natl. Acad. Sci. USA 77, 2415-2418 (1980).

23. Ebels, I.: J. Neural. Transmission 49, 87-105 (1980).

24. L'Helias, C., Ebels, I., Salemink, C.A.: J. Insect Physiol. 28, 459-463 (1982).

25. Rembold, H., Buff, K., Hennings, G.: Clinica Chimica Acta 76, 329-338 (1977).

26. Ziegler, I., Maier, K., Fink, M.: Cancer Research 42, 1567-1573 (1982).

27. Fink, M., Ziegler, I., Maier, K., Wilmanns, W.: Cancer Research 42, 1574-1578 (1982).

28. Rembold, H., Vaubel, A.: Z. physiol. Chem. 351, 1277-1279 (1970).

29. Rembold, H., Langenbach, T.: J. Protozool. 25, 404-408 (1978).

30. Rembold, H., Metzger, H.: Z. Naturforsch. 22, 827-830 (1967).

31. Blair, J.A., Ratanasthien, K., Leeming, R.J.: Gut 15, 832 (1974).

32. Kettler, R., Bartholini, G., Pletscher, A.: Nature 249, 476-478 (1974).

33. Rembold, H., Vaubel, A., Rao, P.J.: Arch. Microbiol. 97, 51-62 (1974).

34. Häberle, D.A., Schiffl, H., Mayer, G., Hennings, G., Rembold, H.: Pflügers Arch. 375, 9-16 (1978).

35. Häberle, D.A., Schiffl, H., Mayer, G., Davis, J.M., Hennings, G., Rembold, H.: Naunyn-Schmiedeberg's Arch. Pharmacol. 306, 287-293 (1979).

36. Hennings, G.: Dissertation Univ. München 1977, p. 64.

37. Rokos, H., Rokos, K., Frisius, H., Kirstaedter, H.-J.: Clinica Chimica Acta 105, 275-286 (1980).

38. Rembold, H., Metzger, H., Sudershan, P., Gutensohn, W.: Biochim. Biophys. Acta 184, 386-396 (1969).

39. Rembold, H., Simmersbach, F.: Biochim. Biophys. Acta 184, 589-596 (1969).

40. Rembold, H.: In "Chemistry and Biology of Pteridines" (Iwai, K., ed.), pp. 163-178, Internat. Acad. Printing Co., Tokyo, 1970.

41. Rembold, H., Metzger, H., Gutensohn, W.: Biochim. Biophys. Acta 230, 117-126 (1971).

42. Rembold, H., Chandrashekar, V., Sudershan, P.: Biochim. Biophys. Acta 237, 365-368 (1971).

43. Gutensohn, W.: Dissertation Univ. München, 1968.

44. Rembold, H., Gutensohn, W.: Biochem. Biophys. Res. Commun. 31, 837-841 (1968).

45. Wachter, H., Grassmayr, K., Hausen, A.: Cancer Letters 6, 61-66 (1979).

46. Wachter, H., Hausen, A., Grassmayr, K.: Z. physiol. Chem. 360, 1957-1960 (1979).

47. Danks, D.M., Schlesinger, P., Firgaira, F., Cotton, R.G. H., Watson, B.M., Rembold, H., Hennings, G.: Pediat. Res. 13, 1150-1155 (1979).

48. Watson, B.M., Schlesinger, P., Cotton, R.G.H.: Clin. Chim. Acta 78, 417-423 (1977).

2

QUANTITATIVE CHEMICAL ANALYSIS OF PTERIDINES

AUTOMATIZED ROUTINE ESTIMATION OF NEOPTERIN IN HUMAN URINE BY HPLC ON REVERSED PHASE

Dietmar Fuchs, Arno Hausen, Gilbert Reibnegger, Helmut Wachter
Institute of Medical Chemistry and Biochemistry, University of Innsbruck, Fritz-Pregl-Straße 3, A-6020 Innsbruck, Austria.

Summary

The presented HPLC-estimation on reversed phase is optimized to quantify neopterin, creatinine and uric acid. Eluted neopterin is monitored by its fluorescence (353 nm excitation, 438 nm emission), creatinine and uric acid by their absorption at 235 nm. The stability of urinary neopterin samples was controlled during storage and UV-irradiation. Daily neopterin excretion of healthy individuals was estimated by this method. The neopterin values were advantageously related to bodysize when comparing various persons. A correlation between neopterin excretion and daily rhythm was demonstrated. Regenerative phases like sleep time were accompanied by the highest neopterin values whereas the lowest values were found in stress situations.

Introduction

Increased neopterin excretion was found in patients with various types of malignancies or virus diseases (1). In order to study the relation between malignancies and elevated neopterin levels we examined the significance of urinary neopterin in the detection and monitoring treatment of carcinomas in homogenous groups of patients (2,3). Therefore, a simple and rapid method for measuring urinary pteridines is necessi-

Biochemical and Clinical Aspects of Pteridines · Volume 1
Cancer · Immunology · Metabolic Diseases

tated, which is in agreement with the required statistical quality concerning specificity, accuracy and precision. The presented HPLC-method complies with these conditions optimized to simultaneously estimate neopterin, creatinine and uric acid.

Methods

Sample preparation

To dissolve possible urinary sediments 200 µl of urine were mixed with 1 ml Sørensen-phosphate buffer 15 mmol/l, pH = 7,0 (= 58.7% $KH_2PO_4 \cdot 3H_2O$; 0.54 mmol/l EDTA = disodium diaminoethylenetetraacetate are added). Up to 16% of neopterin and up to 10% of creatinine were found coprecipitated in sediments of uric acid or sodium urate (4). Samples were either measured immediately or stored at -20°C in the dark. To protect neopterin from photooxidation all operations were done in dim light.

Chromatography

Because unpurified urinary samples were used a guard column was necessary to avoid contamination of the main column (5). We used guard columns (40 x 4 mm I.D.) packed with 10 µm reversed phase C-18 material. After chromatography of about 70 diluted urinary samples cleaning of the guard column is necessary, e.g. using 60% methanol or 30% acetonitrile. Under these conditions one guard column survives at least 500 analyses. Chromatographic separation is done with a ready-to-use column (300 x 3.9 mm I.D.) packed with 10 µm reverse phase C-18 material. A 10 ul diluted urinary sample is injected. Elution is performed with degassed Sørensen-phosphate buffer

15 mmol/l,pH = 6.4 (= 26.6% $KH_2PO_4 \cdot 3H_2O$; temperature of column 30°C). During the first 7.5 min. a flow rate of 1.0 ml/min. is used to separate urinary pteridines. After elution of neopterin during the following minute flow rate is linearly increased to 2.0 ml/min. and kept constant for 14 minutes. Thus, fast elution of less polar compounds is reached; measurement of which is not of interest in this study. In the following minute, flow is reduced to starting conditions (1 ml/min.). After 15 minutes the next analysis can be performed. It is necessary to control quantitation with a standard sample after measuring 7 urinary samples. Seventy analyses can be done daily including sample preparation, quantitation of external standards and purifying of columns. A brief summary of this chromatographic procedure is given in table 1.

Detection

Neopterin is detected by its fluorescence at an excitation of 353 nm and emission of 438 nm. Creatinine and uric acid are both monitored by their absorption at 235 nm. Chromatograms are electronically integrated and quantified using the external standard method. In figure 1 a typical chromatogram of a urinary sample is shown. The chromatographic method is totally automatized: autosampler, pump, injection valve, column temperature and integrators are controlled by a data system.

Linearity and sensitivity

The sensitivity at a peak-noise ratio of 5:1 was determined for neopterin to be 72 nmol/l of undiluted sample or 120 fmol per injection, for creatinine and uric acid these limits are 36 pmol per injection and 50 pmol, respectively. The upper limits of linearity were found to be 40 umol/l for neopterin, 26 mmol/l for creatinine and 6 mmol/l for uric acid (4,6,7).

Table 1: Quantitative estimation of neopterin and creatinine by HPLC

Sample preparation:	200 µl urine diluted with Sørensen-phosphate buffer (pH = 7.0; 15 mmol/l), inclusive 0.54 mmol/l EDTA
Chromatograph:	Model LC 5000, Fa. Varian, Palo Alto, Ca, USA.
Autosampler:	Model LC 8000, Fa. Varian, Palo Alto, Ca, USA.
Injection:	Air actuated loop injector, Fa. Valco Instruments, Houston, Tx, USA
Guard column:	Fa. Varian, packed with 10 µm reversed phase C-18-ODs 2, Fa. Whatman, Clifton, NJ, USA
Main column:	Stainless steel, 10 µm Bondapak C-18, Fa. Waters, Milford, Ma, USA
Elution:	Degassed Sørensen-phosphate buffer, pH = 6.4, 15 mmol/l 0 - 7.5 min. flow is 1.0 ml/min 7.5 - 8.5 min. flow gradient 1.0 - 2.0 ml/min. 8.5 - 14 min. flow is 2.0 ml/min 14 - 15 min. flow gradient 2.0 - 1.0 ml/min.
Detection:	Spectrophotometer SFM 23, Fa. Kontron, Eching-Munich, FRG, neopterin: 353 nm excitation, 438 nm emission. UV-absorption detector PM2-DLC, Fa. Zeiss, Oberkochen, FRG, creatinine: 235 nm
Evaluation:	Chromatopac CR-1A, Shimadzu Coorp., Kyoto, Japan; CDS 111 LC, Fa. Varian.
Control system:	CDS 111 LC, Fa. Varian.

Precision and accuracy

Precision and accuracy of this HPLC-method were estimated measuring neopterin levels simultaneously with creatinine levels; within-run precision was obtained on three urinary samples measured 20 or 24 times; day-to-day precision was observed also using three urinary samples during 15 days. Data are shown in table 2. A mean value of 99.3% for neopterin/creatinine ratio was found in recovery experiments of six spiked urine samples.

Fig. 1 Chromatogram of a urinary sample chromatographed under the described conditions

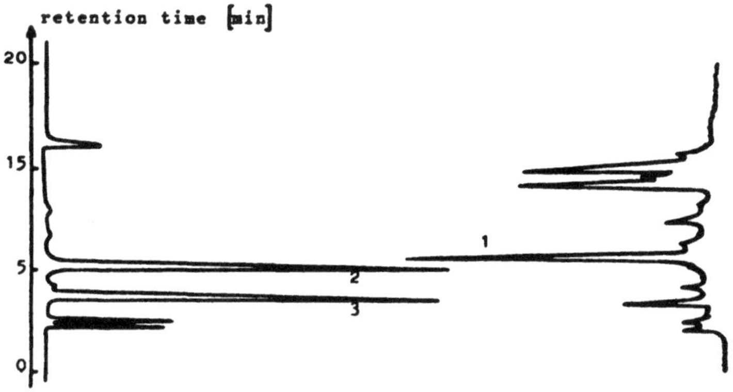

left chromatogram was obtained with the UV - detector (235 nm), the right one with the fluorescence detector (353 nm excitation, 438 nm measuring); peaks: 1 = neopterin, 2 = creatinine, 3 = uric acid

Table 2: Precision of urinary neopterin in relation to creatinine determined by HPLC on reversed phase

sample	1	2	3	1	2	3
n	20	24	24	15	15	15
x (µmol/mol)	95	214	244	183	260	229
C.V. (%)	6.5	3.8	3.8	7.4	2.8	7.1
	within-run			between-day		

Stability of urinary neopterin

Stability of neopterin was tested on undiluted urinary samples and standard samples (Figures 2, 3 and table 3).
When exposed to daylight in closed bottles at room temperature (25°C), neopterin was destroyed within a few hours due to photooxidation; it was almost totally oxidized to pterin-6-carbonic acid.

Fig.2 Stability of neopterin in standard and urinary samples

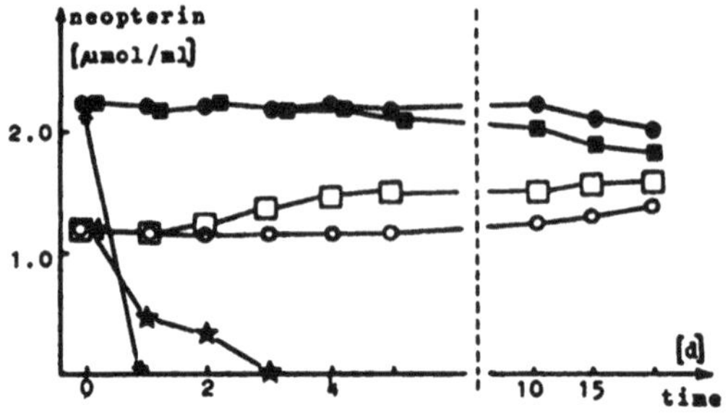

$\bar{x}$ of 5 parallel experiments is shown
stored in the dark (4°C): ● standard, ○ urine
stored in the dark (27°C): ■ standard, □ urine
stored at daylight (27°C): ● standard, ★ urine

When urinary samples were kept in closed brown bottles in dim light and at room temperature an increase of neopterin concentration was found after 3 days, which may descend from reduced neopterin (7,8-dihydroneopterin, 5,6,7,8-tetrahydroneopterin). Addition of ascorbic acid seems to catalyse this oxidation process (Table 3). Addition of 10 mg Dithioerythritol pro ml had no influence.

Fig.3 Photooxidation of neopterin in dependence of time

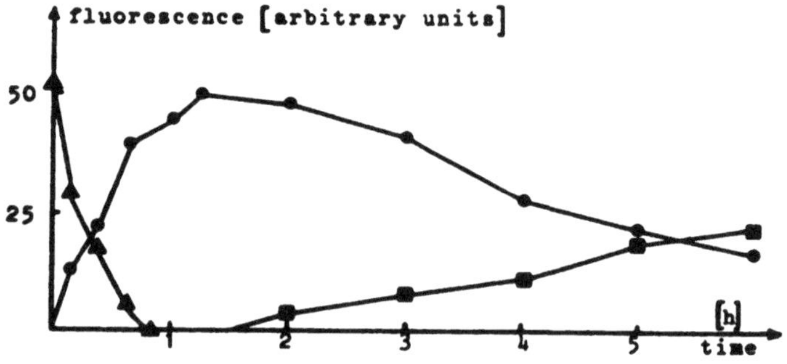

standard samples were irradiated at 366nm (Uvis mercury lamp, Fa. Desaga, Heidelberg, G.F.R.)
▲ neopterin, ● pterin 6 carbonic acid, ■ pterin

Table 3: Influence of ascorbic acid to urinary neopterin/creatinine ratio (100 mg of ascorbic acid added to 10 ml of three fresh urinary samples; storage at room temperature in the dark).

day of analysis	0	1	3	5	7	9
$\bar{x}$ (μmol/mol)	141	180	184	197	223	225
increase (%)	0	27	31	40	58	60

At 4°C in the dark after about two weeks neopterin values still remain unchanged, then neopterin contents increased very slightly. Best stability was provided when urinary samples were stored at -20°C in the dark. After 5 months there was no significant difference of neopterin/creatinine ratio compared with starting values (Table 4).

Table 4: Neopterin/creatinine ratio (μmol/mol) of urines stored for 5 months at -20°C in the dark compared with the starting values

n	$\bar{x}$ start	$\bar{x}$ after storage	$\frac{\bar{x}\text{ start}}{\bar{x}\text{ storage}} \cdot 100$
10	255	274	107.5%

Influence of irradiation on neopterin standard samples is shown in more detail in figure 3. Standard samples were exposed to UV-irradiation, marked photooxidation was seen within 5 minutes. Thus, if urinary samples are not quantified immediately, they should be kept frozen at -20°C in the dark.

Normal values - Neopterin/creatinine ratio

In order to compensate physiological variations of urinary concentration it is useful to relate urinary neopterin to creatinine (4). Urinary neopterin levels of 417 healthy subjects were estimated in their first morning urine. We found a biological variation of about 30% (Table 5).

Table 5: Urinary neopterin of healthy adults

sex	n	age (years)	$\bar{x}$ (neopterin/creatinine) (μmol/mol)	S.D.	C.V.%
♂	208	19-80	113 ±	35.3	31.2
♀	209	17-85	140 ±	35.7	25.5

There is a sex and age dependence in the neopterin/creatinine ratio (4) which may mainly be influenced by creatinine-variability. Similar coefficients of variation were obtained for individuals observed during a one month period (4).

Daily excreted urinary neopterin

Daily excretion of neopterin investigated in 24 hour urines was found to be quite constant. Since excreted creatinine is reported to be dependent on muscle mass (8) it is necessary to consider body structure when comparing daily neopterin excretion of different subjects. Body size seems to be the best parameter and is better than a relationship to body surface or body weight. Statistical utility of daily neopterin excretion related to body size is similar to the neopterin/creatinine ratio estimated in first morning urine (Table 6).

Table 6: Daily excretion of neopterin in healthy individuals

	neopterin / 24 hours (µmol/d)	neopterin / size, day (µmol/d,cm)	neopterin / creatinine (µmol/mol)	age (years)
n	22	22	22	18
$\bar{x}$	1.58	8.71	114	-
C.V.%	31.8	23.8	21.4	51
n	13	13	13	18
$\bar{x}$	1.23	7.94	131	-
C.V.%	28.4	21.7	22.6	37

Thus estimation of creatinine additional to neopterin values is not absolutely essential if 24 hour urine is used. Calculation of daily neopterin excretion using various spontaneous urines with formular given below leads to practicable results.

$$\frac{\text{neopterin}}{\text{hour}} = \frac{\text{neopterin}}{\text{ml}} \cdot \frac{\text{volume of urine [ml]}}{\text{interval (measured urine - last urine before [h])}}$$

However, routine calculation in this way does not seem to be very accurate, because volume and interval of urine collections by patients were not carried out exactly.

Figure 4 depicts excretion of neopterin per hour related to bodysize (cm) in relation to the time of day. 9 individuals were watched at different times and neopterin excretion per hour was calculated from spontaneous urines (table 7).There is a higher neopterin excretion or production during sleep or during regenerative phase, respectively. In addition, during stress situations a strong decrease of urinary neopterin levels was

Table 7: Neopterin excretion per hour of one healthy person calculated from 58 spontaneous urines collected over a one month period

	Creatinine	Neopterin	Neopterin / Creatinine	Creatinine	Neopterin
	mmol/l	nmol/l	µmol/mol	mmol/d	nmol/d
$\bar{x}$	15.1	1.59	102	17.4	1.73
S.D.	4.87	0.70	23.6	2.80	0.35
C.V.%	32.3	44.0	23.1	16.1	20.2

seen (data not shown). This observation could be the result of raised adrenalin levels. Biosynthesis of adrenalin and nor-adrenalin is enzymatically dependent on tetrahydrobiopterin as cofactor (9). An increased consumption of dihydroneopterinintriphosphate to produce tetrahydrobiopterin may lead to lowered urinary neopterin levels. Neopterin pro creatinine

Fig.4 Neopterin excretion per hour and per bodysize

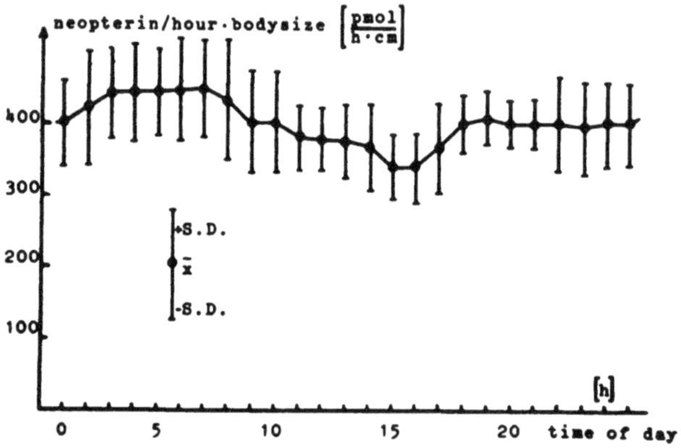

excretion of neopterin per hour was calculated from intervals between urinary samples, 5 individuals were watched for three days

TABLE 8: Daily variation of neopterin/creatinine ratio (µmol/mol)

Time	7.00	12.00	16.00	19.00
n	17	19	17	19
$\bar{x}$	149	124	110	112
S.D.	32	30	29	29
C.V.%	21.5	24.2	26.4	25.9

ratios show the same dependence (Table 8). Due to these findings, one has to compare urine, if possible, from the same time of day when using spontaneous urine for clinical routine estimations. First morning urine is the most suitable one, because it shows the smallest variations in neopterin/creatinine ratio.

Determination of 7,8-dihydroneopterin

It is our intention to estimate 7,8-dihydroneopterin in order to test if high urinary neopterin levels are accompanied by increased levels of the reduced forms. Oxidation of hydrogenated neopterin did not lead to satisfying results (4). We therefore searched for a way to quantity 7,8-dihydroneopterin directly without an oxidation step during sample preparation. Detection of reduced forms of neopterin is possible electrochemically following HPLC.

The low oxidation potential, which is sufficient for the estimation of 7.8 dihydroneopterin, may lead to high selectivity. Pilot experiments concerning this estimation are encouraging. Investigations to determine dihydroneopterin concentration in human urine are in progress.

Conclusion

This HPLC method was developed in order to allow estimation of neopterin values in a greater number of urinary samples and to realize detailed clinical studies. Using this simple, practicable and exact method accomplishment of this desire is possible with minimal expense. The costs of this method are very low, if a HPLC system is on hand.

The primary clinical investigations are also published in this book. Basic considerations on normal values showed that one can analyse spontaneous urines. In this case it is necessary to relate neopterin to creatinine to compensate physiological variations of urinary concentration. Creatinine can be determined in one single run simultaneously by this HPLC method. It ought to be mentioned that urine should be collected at uniform times; the best comparable results are given by first morning urine since the sleeping period before urine collection leads to the statistically best neopterin mean values when determined for healthy individuals. If daily neopterin excretion is known, the evaluation of creatinine concentration is not essential, but the collection of 24-hour-urine means an additional expense for patient and physician. Neopterin values must be related to body size or a similar parameter. We found acceptable results, comparable to neopterin pro creatinine ratio of spontaneous urines.

Because of photosensibility of pteridines storage and operations should be carried out in darkened rooms. If samples cannot be measured immediately they should be stored at -20 $^{\circ}$C in the dark. To elucidate the connection of urinary neopterin with 7,8-dihydroneopterin levels we tried to find a method to estimate reduced pteridines directly without an oxidation step during sample preparation. First results of electrochemical oxidation of 7,8-dihydroneopterin were promising.

References

1. Wachter, H., Hausen, A. and Grassmayr, K.
Hoppe Seyler's Z. Physiol. Chem. 360, 1957-1960(1979).

2. Hausen, A., Fuchs, D., Grünewald, K., Huber, H., König, K. and Wachter, H.
Clin. Chem. Acta 117, 297-305(1981).

3. Bichler, A., Fuchs, D., Hausen, A., Hetzel, H., König, K. and Wachter, H.
Clin. Biochem. 15 (1),38-40(1982)

4. Hausen, A., Fuchs, D., König, K and Wachter, H.
J. Chromatogr., 227, 61-70 (1982)

5. Hausen, A., Fuchs, D., Grünewald, K., Huber, H., König, K. and Wachter, H.
Clin. Biochem. 15 (1),34-37(1982)

6. Hausen, A., Fuchs, D. and Wachter, H.
J. Clin. Chem. Clin. Biochem. 19, 375-378 (1981)

7. Hausen, A., Fuchs, D., König, K., Wachter, H.
Clin.Chem. 27, 1455 -1456 (1981)

8. Schöch, G., Garbrecht, M., Heller-Schöch, G., Baisch, H. Leifer, W.
Blut 38, 391-396 (1979).

9. Kaufmann, S.
J. Biol. Chem. 242 (17), 3934 (1967).

Acknowledgement

We thank Prof. Dr. W. Pfleiderer and Prof. Dr. A. Niederwieser for their kind gifts of 7.8-dihydroneopterin and 5,6,7,8-tetrahydroneopterin.

AUTOMATIC HPLC OF PTERINS WITH OR WITHOUT COLUMN SWITCHING

A. Niederwieser, W. Staudenmann and E. Wetzel

Division of Clinical Chemistry, Dept of Pediatrics, University of Zurich, Steinwiesstr. 75, CH-8032 Zurich, Switzerland

An automatic system with facilities of column switching allows the omission of sample pretreatment for reverse phase HPLC analysis of total pterins in urine (biopterin, neopterin, monapterin, isoxanthopterin, pterin, 6-hydroxymethylpterin). A fast and quantitative method for oxidation of reduced forms of pterins with manganese dioxide is described. Several HPLC variants have been developed including programs for measuring the sepia analogs and of tetrahydrobiopterin. The latter is resolved into the 6-R,S enantiomers and monitored amperometrically.

Introduction

Bailey and Ayling (1,2) and Stea *et al.* (3,4) were the first groups, who described HPLC of unconjugated pterins. They used cation-exchange materials like Partisil SCX as the stationary phase. At that time we became interested in separation and quantitation of pterins in urine for the study of cofactor-deficient variants of hyperphenylalaninemia. High voltage electrophoresis (5) and paper chromatography (6) were useful to us (7), but allowed only semiquantitative measurements. Gas chromatography-mass fragmentography (8) gave valuable results (9), but was expensive and rather tedious. Trimethylsilyl derivatives, particularly of neopterin and biopterin, exhibited adsorption effects on glass capillary columns and in the GC-MS interface; nonlinear calibration curves were obtained with different slopes for pure reference solutions and urine, respectively. These

Biochemical and Clinical Aspects of Pteridines · Volume 1
Cancer · Immunology · Metabolic Diseases

problems led us to abandon this selective technique and to employ HPLC. In our hands, ion-exchange HPLC (1-4) gave non-Gaussian peaks with disturbing tailing and overlaping of neopterin and monapterin. Better results were obtained by reverse phase HPLC on LiChrosorb RP-8 (10). In the meantime, reverse phase HPLC with many helpful technical details have been described by Fukushima and Nixon (11) for the analysis of reduced forms of biopterin and other pterins, in biological tissues and fluids, by Kohashi *et al.* (12) for the analysis in food and human urine, and by others for the analysis of urine (13-15).

Our aim was to overcome the time consuming prepurification steps which are lowering yield and precision, and to develop a reliable routine method. The method should allow not only measurement of total biopterin and neopterin to screen for tetrahydrobiopterin deficiency, but also detection of unusual pterins - if any - of quite different polarity. Such a pterin might be expected to accumulate in a new variant of hyperphenylalaninemia.

In the following, we describe an automatic system with facilities of column switching. The method requires no pretreatment of sample except for a simple oxidation step. In addition, special programs are described for analysis of yellow fluorescing pterins in native urine and of tetrahydrobiopterin.

Materials

Chemicals

Biopterin, neopterin, both reduced and oxidized, 2'-deoxysepiapterin, 3'-hydroxysepiapterin, sepiapterin (Dr. B. Schircks, Wettswil a.A., Switzerland); 6,7-dimethylpterin, 6-formylpterin, monapterin=L-threo-neopterin, 6- and 7-monomethylpterin, (6S)- and (6R)-tetrahydrobiopterin·2HCl enantiomers (gifts of Prof. M. Viscontini, University of Zurich, Switzerland); 6-hydroxymethyl pterin (gift of Prof. M. Pfleiderer, University of Constance, FRG); isoxanthopterin, pterin, pterin-6-carbonic acid, riboflavin,xanthopterin (Fluka AG, Buchs SG, Switzerland); 6-phenylpterin (gift of Dr. B. Levine, NIH, Bethesda, MD USA).

Florisil 60-100 mesh, LiChrosorb RP-8 and RP-18, both 5 µm, Lewatit MP 5080, 100-200 mesh, manganese dioxide (Merck AG, Darmstadt, FRG).

HPLC equipment
Empty columns 40 mm no. 103.13 and 120 mm no. 103.03 (Knauer, Berlin FRG); HPLC column 250x4.9 mm Spherisorb S5 ODS no. 254-101, 3 pumps Model 410, mixing chamber, programmer Model 200, 1 three-way valve RE 5302 with pneumatic device RE 5300 (Kontron, Zurich, Switzerland); injector Model U6K, automatic sampler WISP 710 (Waters Assoc., Milford, MA,USA); 3 six-port valves Valco AH-CV-6-UHPa-N60 (Valco Instruments, Houston, TX, USA); fluorometer Model 650-10S (Perkin Elmer, Oak Brook, IL, USA); filter fluorometer Schoeffel FS 950 with lamp FSA 110 and filter FSA 403 (Kratos GmbH, Karlsruhe, FRG) and emission filter B-40: 447 nm with 20 nm band-width (Balzers, Principality of Liechtenstein); integrator HP 3390 A (Hewlett Packard, Avondale, USA); amperometric detector LC4A with flow cell LC17 glassy carbon (Bioanalytical Systems Inc., West Lafayette, IN, USA); pump Model 110A with pulse damping device (Altex, Berkeley, CA, USA).

External standard mixture, blue fluorescing
Neopterin, monapterin, biopterin and pterin, each 0.5 mg/l, and isoxanthopterin, 0.2 mg/l, in 0.1 M HCl.

External standard mixture, yellow fluorescing
3'-Hydroxysepiapterin, sepiapterin, 2'-deoxysepiapterin, each 2.0 mg/l, and xanthopterin, 2.0 mg/l, in H_2O. Stored frozen in small vials, under inert gas.

Internal standard, blue fluorescing
6-Hydroxymethyl pterin, 30 mg/l in 0.1 M HCl. Alternatively, 6-methylpterin, 30 mg/l in 0.1 M HCl can be used.

Tetrahydrobiopterin stock solution
Tetrahydrobiopterin dihydrochloride, 0.1 g/l and dithiothreitol, 1 g/l, dissolved in deaerated 0.1 M HCl. Stored frozen in small vials under inert gas. This stock solution is diluted with the deaerated 0.1 M HCl just before use.

Solvents for HPLC
A : Methanol-water (3:97, v/v)
B : Isopropanol-methanol-acetic acid (49:49:2, v/v)
C : Isopropanol-methanol-water (1:1:8, v/v)
D : Isopropanol-water (1:1, v/v)
E : Isopropanol-methanol-acetic acid-water (5:5:0.5:990, v/v)
F : KH_2PO_4, 100 µmol/l, adjusted to pH 4 with citric acid, EDTA, 60 µmol/l, 2% v/v methanol.

Methods

Depending on the HPLC equipment available and the kind of pterins to be analyzed, several HPLC variants can be used. An overview is given in Table 1.

Oxidation for HPLC with column switching
Reduced forms of pterins were oxidized for stabilization and to simplify the analytical system. Total biopterin, neopterin, etc. can be measured in this way.

The frozen urine specimen is thawed in a microwave oven, mixed by ultrasonication, and an aliquot of 500 µl is transferred

TABLE 1

Summary of HPLC variants described for analysis of blue and yellow fluorescing pterins and tetrahydrobiopterin. + necessary, - not necessary.

Pterins analyzed (Fluoresc. nm)	Sample		Standards		HPLC variant	Column switching	Gradient elution	Solvents	Cycle time (min)
	oxidation	pretreatment	extern.	intern.					
total (blue 350/450)	+	-	+		I	+	+	A,B,C	55
	+	+		+	II	-	+	A,B	40
	+	+		+	III	-	-	A,(B,D)	20
sepia-analogs xanthopterin (yellow 425/530)	-	-	+		IV	-	+	B,E	35
tetrahydrobiopterin	-	-	+		V	-	-	F	10

immediately in a 10 ml plastic tube. Acidify to pH 1-1.5 with 6 M HCl (usually 20 µl is sufficient), add approx. 10 mg manganese dioxide and shake the mixture gently for 5 minutes. Then, admix 5 ml of water (or only 2 ml, if creatinine concentration in the urine is less than 2 mmol/l) and centrifuge for 5 min at 4000 r.p.m. Transfer 200 µl of the clear supernatant into a vial for HPLC.

Oxidation and sample pretreatment for HPLC without column switching

Take 1 ml of urine (or 5 ml, if creatinine concentration is less than 1 mmol/l), add 100 µl of internal standard solution and oxidize as above. Apply the clear supernatant (approx. 0.7 or 4 ml, respectively) onto 1 g of Florisil within a column of approx. 9 mm inner diameter, with Florisil previously rinsed with 10 ml 1 M HCl followed by 10 ml of 0.1 M HCl. Then rinse the column with 10 ml 0.1 M HCl and 10 ml of H_2O. Elute the pterins with 2x5 ml acetonitrile-pyridine-water-0.1 M aq. ammonia (60:3:57:4, v/v). Adjust the pH to 10 with 0.1 M NaOH and apply the eluate onto a column (30x9 mm) of Lewatit MP 5080 (acetate form). Rinse with 10 ml of H_2O and elute the pterins with 2x5 ml of 1M aq. formic acid. The sample is then ready for HPLC, and can be concentrated by evaporation to dryness and dissolved in an appropriate volume (sonication!), if necessary.

Fluorometer settings for blue fluorescing pterins

Perkin Elmer Model 650-10S: 350/450 nm, each slit 10 nm, range 0.3, response norm (2 sec).

Schoeffel FS 950: Lamp FSA 110, Ex. filter FSA 403, emission filter B-40 (Balzers) 447 nm, band-width 20 nm, range 0.05, sensitivity 800V, time constant 2 sec.

Fluorometer settings for yellow fluorescing pterins

Perkin Elmer Model 650-10S: 425/530 nm, each slit 15 nm, range 1.0.

Sample storage

Aliquots of urine to be analyzed, 5-10 ml, should be protected from light, frozen in plastic tubes immediately after collection and should be thawed only just before use.

For analysis of yellow fluorescing sepia-analogs, the urine must be bubbled through with nitrogen, shock frozen immediately after collection, and stored at -20°C under inert gas. After thawing, the sample is sonicated for 10-30 sec and centrifuged at 4000 r.p.m. for 5 minutes. Immediately thereafter, 5 µl of the supernatant is injected for HPLC.

I. HPLC with column switching and gradient elution
(No sample pretreatment necessary)

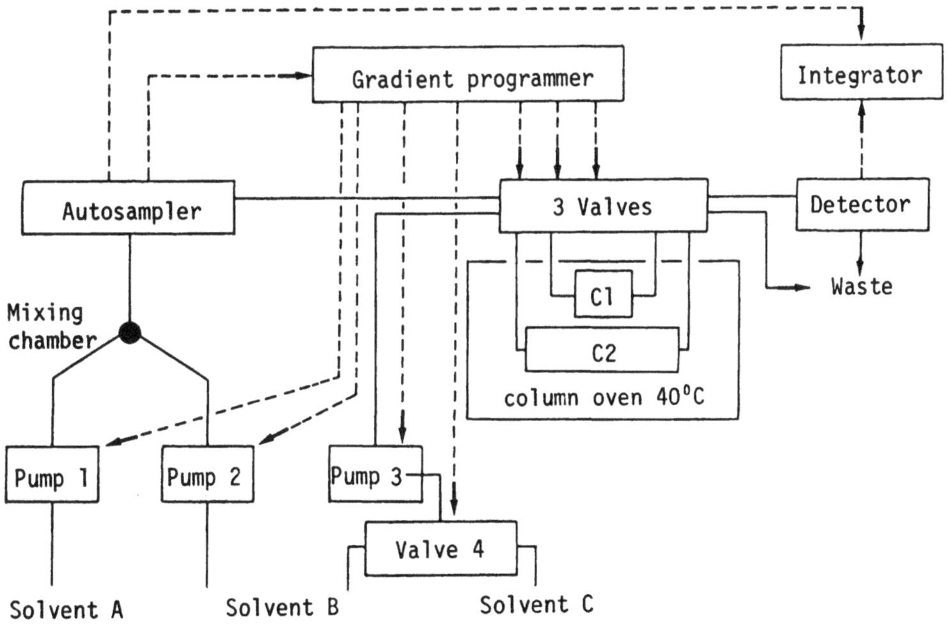

Fig. 1. Equipment needed for HPLC with column switching and gradient elution. Precolumn C1: 40x4.6 mm LiChrosorb RP-8, 5 µm. Analytical column C2: 250x4.9 mm Spherisorb S5 ODS. The autosampler starts the program and the integrator. The gradient programmer also controls the valves and pump 3 which is used for rinsing of the precolumn C1 only. Details of the 3 valves and their different positions, see Fig. 2.

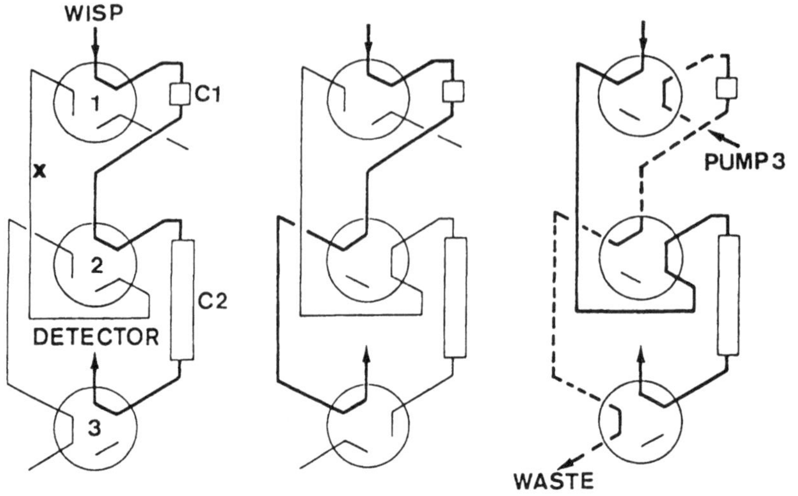

Fig. 2. Connections and different positions of the 3 high-pressure valves used for column switching. First status (a), 0-2 min: Sample injection with columns 1 and 2 in series. Second status (b), 2-12 min: Gradient elution of slow moving compounds from column 1 into detector; column 2 disconnected and stopped. Third status (c), 12-32 min: Elution from column 2 into detector; cleaning of column 1.

The equipment and connections needed are shown in Figs. 1 and 2. The sample is injected onto precolumn C1 connected in series with the analytical column C2, both equilibrated with solvent A (see above). The faster moving compounds till and including pterin are chromatographed with solvent A into C2. After 2 min, before any of the compounds are able to elute from C2, the analytical column is disconnected and stopped temporarily. The slow moving compounds are eluted with a gradient from the precolumn into the detector. Thus, the slow moving compounds appear on the chromatogram first. Then, the precolumn is disconnected and thoroughly rinsed with solvents B and C, twice, for cleaning. During that time, the compounds which were trapped temporarily within the analytical column, are eluted with solvents A and B. After elution of pterin (32 min), both columns are connected in series and equilibrated again. Details of the program, see Table 2.

TABLE 2

Program for HPLC with column switching and gradient elution. The auxiliary functions (flags) Aux 1,2,3 switch the valves 1,2,3, respectively (see Fig. 2). Aux 4 switches on pump 3 operating at a flow rate of 1 ml/min; Aux 5 switches on valve 4 for solvent C (at off position solvent B is running, see Fig. 1, as long as pump 3 is on).

Time (min)	Function	duration (min)	Remarks
0	0% solvent B		solvent A is running
	flow 1.3 ml/min		
2	Aux 2	30	disconnect column 2
	Aux 3	10	connect column 1 with detector for 10 min
2.1	5% solvent B		
4.0	20% solvent B	5	5 → 20% B within 5 min
9.0	0% solvent B		
	flow 2 ml/min		
12.0	Aux 1	20	connect column 2 with detector for 20 min
	Aux 4	20	start pump 3 for 20 min (1 ml/min)
	Aux 5	5	rinse column 1 with solvent C for 5 min
	flow 1.2 ml/min		
19.0	20% solvent B	10	0 → 20% B within 10 min
22.0	Aux 5	5	rinse column 1 with solvent C for 5 min
29.0	100% solvent B		
32.0	0% solvent B		equilibrate with solvent A
53.0	Aux 1	1	rinse capillary X (Fig. 2) with solvent A
55.0	flow 1.3 ml/min		next sample injected

II. HPLC with gradient elution (no column switching)

Less equipment is needed for this technique, as shown in Fig. 3.

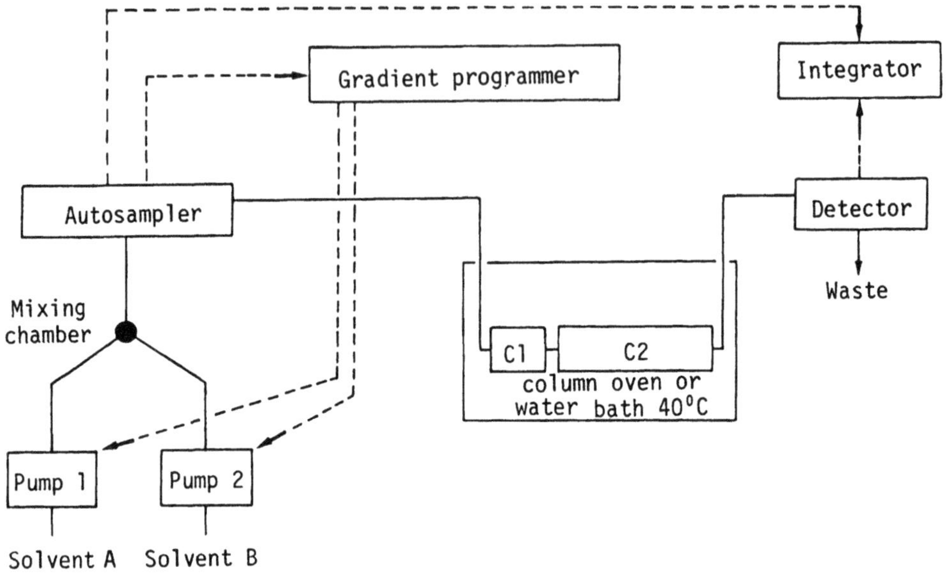

Fig. 3. Equipment needed for HPLC with gradient elution (without column switching). Precolumn C1 and analytical column C2, see legend to Fig. 1.

However, pretreatment of biological samples is necessary (see above). In order to allow analysis of large series of samples without interruption, the columns are rinsed after each analysis automatically. The program including re-equilibration of the stationary phase lasts 40 min (Table 3).

III. HPLC with isocratic elution (no column switching)

This simple version needs only a pump, an injector (e.g. Waters U6K), two columns (guard and analytical) and a detector. Naturally, biological samples have to be pretreated (see above). Oxidized pterins are eluted isocratically with solvent A (see above) at a flow rate of 1 ml/min or more. After 20 min, when pterin has eluted, the next sample can be injected. After

TABLE 3

Program for HPLC with gradient elution (without column switching).

Time (min)	Function	Remarks
0	0% solvent B, flow 1.0 ml/min	solvent A is running, column 1 and 2 in series
10	20% solvent B, duration 10 min	0 → 20% B within 10 min
20	100% solvent B	
23	0% solvent B	equilibration with solvent A
40		next sample injected

10-20 injections of biological samples, the columns have to be rinsed with solvent D followed by solvent B, to regain resolution of isoxanthopterin and biopterin.

IV. HPLC of yellow fluorescing pterins

Sepiapterin, 3'-hydroxysepiapterin, 2'-deoxysepiapterin, etc. should be analyzed only in samples stored under inert gas (see above). They are analyzed best with a special program which needs a gradient system, an injector (by hand: U6K or an autosampler, e.g. WISP), a 120x4.6 mm column LiChrosorb RP-8, 5 µm, equilibrated with solvent E and fluorimetric setting at 425/530 nm (see above). The program is listed in Table 4. The column must be cleaned after approx. 20 injections of urine samples.

V. HPLC of tetrahydrobiopterin

Tetrahydro pterins are very unstable against oxygen. Analysis of the native forms needs stabilization of the sample (see above). Standard solutions with 0.1 M HCl have to be made fresh by dilution of the stock solution (see above). HPLC is performed

TABLE 4

Program for HPLC of yellow fluorescing pterins.

Time (min)	Function	Duration min	Flow rate ml/min	Remarks
0	0% solvent B		1	solvent E is running
2	2% solvent B			
9	20% solvent B	10		2 → 20% B within 10 min
20	100% solvent B			
23	0% solvent B			
25	100% solvent B			
27	0% solvent B		2	re-equilibration with solvent E
35				next sample injected

on a 40x4.6 mm LiChrosorb RP-18, 5 μm precolumn connected in series with a 250x4.9 mm Spherisorb S5 ODS analytical column. Solvent F (see above) is used as the mobile phase; it must be deaerated and kept under inert gas. Detection is best amperometrically and +200 mV, range 0.2 μA full scale and 5 sec time constant. Tetrahydro and dihydro biopterins do not fluoresce at all.

Calculation

When the analysis was performed with external standards only, the following formula was used for calculations:

$$C_{sample} = R \cdot C_{ext.\,standard} \cdot D$$

with C: concentration, R: ratio of peak area in sample/standard, D: dilution factor of sample during oxidation, e.g. D = 11 (0.5 ml urine + 5 ml H_2O) or D = 5 (0.5 ml urine + 2.5 ml H_2O).

When pretreatment of sample and addition of an internal standard was necessary, calculation was performed using the formula:

$$C_{sample} = R \cdot C_{ext.standard} \cdot \frac{R_{int.standard}}{V_{sample}}$$

whereby $R_{int.standard}$: peak area of internal standard expected (pure internal standard solution analyzed after adequate dilution, e.g. 100 µl in 10 ml H_2O) divided by peak area of internal standard found in the sample; V_{sample}: sample volume used.

Storage of HPLC columns

The HPLC columns were always stored in solvent B at room temperature to avoid bacterial and fungal growth. They were tightly closed to avoid desiccation.

Cleaning of HPLC columns

When chromatographic resolution was lost and/or a high back pressure was built up, the column could be cleaned and reactivated by rinsing sequentially as follows:
20-50 ml of solvent B, 70-100 ml of chloroform; 20-50 ml of solvent B, 100 ml of 1 M pyridinium acetate pH 5, approx. 500 ml of solvent C (overnight), 50-70 ml of 5% aq. acetic acid, then equilibration with solvent A. A"dead volume" at the top of the column (shoulder, broadening of all peaks) could be corrected by filling up with the corresponding adsorbent material.

Results and Discussion

HPLC of blue fluorescing pterins

Chromatograms of a standard mixture of blue fluorescing pterins (and pyridoxic acid) obtained with HPLC variants I, II and III are shown in Fig. 4. In variant I, with column switching, compounds moving slower than pterin were eluted first. They

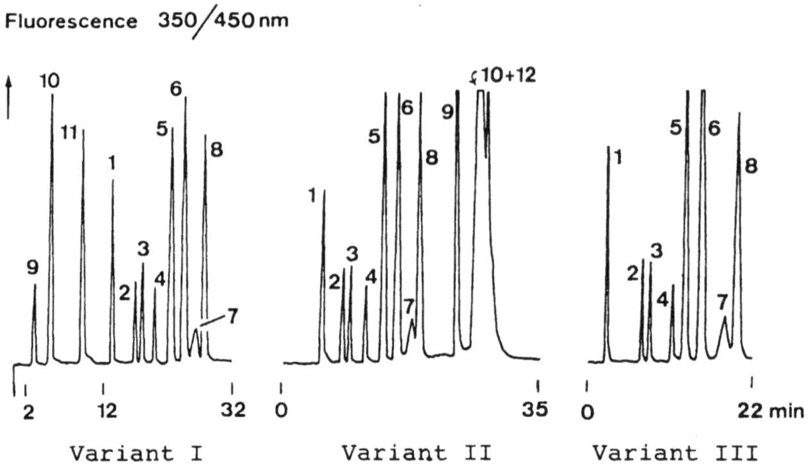

Fig. 4. Reverse phase chromatograms of blue fluorescing pterins with HPLC variant I (column switching and gradient elution), variant II (gradient elution) and variant III (isocratic elution). Peak identification, see Table 5. The amounts injected varied in this mixture from 0.8 (no. 2) to 6 ng (no. 6).

were eluted with a solvent gradient directly from the short precolumn into the detector. Even nonpolar pterins like 6-phenylpterin appeared as sharp peaks within a short period of time, between 2 and 12 minutes. The other pterins of significance in human urine were all separated afterwards. So far, pterin-6-carbonic acid and 6-formylpterin could not be detected in fresh human urine. 6-Formylpterin was found only in samples which had not been protected from light. Typical chromatograms of human urine obtained with HPLC variant I for differential diagnosis of tetrahydrobiopterin deficiency are shown in Figs. 1 and 2 in Chapter 6 in this book. In HPLC variant I, cleaning of the analytical column became necessary only after 200-300 injections of urine samples (without pretreatment). After cleaning with the procedure outlined above, the same column could be used further. For reasons of safety, the precolumn was replaced after approx. 100 injections.

The HPLC variant II gave practically the same chromatogram of fast moving pterins (till pterin, no. 8 in Fig. 4). The solvent gradient enabled recognition of the presence of non-polar pterins. Larger series of samples could only be analyzed by variant II without interruption, if samples had been cleaned by pretreatment.

The simple HPLC variant III could be used without interruption only for 10-20 sample injections. Then the column had to be rinsed with solvents D followed by B. Slow moving urinary components like pyridoxic acid which was found in some samples in higher concentration, led to drifts of the base line. Nevertheless, this variant enabled analysis of small series of samples for biopterin and neopterin, the most important components for differential diagnosis of tetrahydrobiopterin deficiency, with a cycle time of only approx. 20 minutes. Retention times of some of the more important pterins are compiled in Table 5.

The oxidation conditions have been controlled with respect to reaction time, pH and amount of oxidizing reagent added. Aliquots of urine were reacted 5, 10 and 15 minutes with the other conditions as described above. The values of biopterin and neopterin did not differ significantly. Oxidation of aliquots of the urine was also performed at pH 1,5, 1.2 and 0.9. The corresponding values of biopterin and neopterin observed did not differ significantly. The same was true, when 10, 20 or 50 mg of manganese dioxide were used per 500 µl of urine at pH 1.5 or pH 1.0. These results showed that the oxidation conditions are not critical and can be easily reproduced.

To check reproducibility and recovery of the method in biological samples, 3 urine specimen were analyzed without and with the addition of tetrahydrobiopterin, dihydrobiopterin and dihydroneopterin, respectively (Tab. 6). Coefficient of variation was found to be better than ±4% except that of small concen-

TABLE 5

Retention times of blue fluorescing (oxidized) pterins and pyridoxic acid on Spherisorb S5 ODS, 250x4.9 mm in 3 and 30% aqueous methanol, at 40^{0}C.

No.	Compound	Retention time (min) in	
		3% MeOH	30% MeOH
1	Pterin-6-carbonic acid	2.2	
2	Neopterin	7.1	
3	Monapterin	8.1	
4	Isoxanthopterin	11.2	
5	erythro-Biopterin	13.0	4.4
6	6-Hydroxymethylpterin	15.1	
7	6-Formylpterin	18.3*	
8	Pterin	19.9	5.7
9	6-Methylpterin		7.9
	7-Methylpterin		8.6
10	6,7-Dimethylpterin		10.6
11	6-Phenylpterin		35.9
12	Pyridoxic acid		8.8*

*broad peak shape

trations of neopterin. Retrospectively, we found out that this study was performed with suboptimal settings of the integrator and too high sample dilution for neopterin measurement in the low concentration range.

When reduced pterins were added to urine, the recovery of oxidized biopterin and neopterin was higher than 90%, except in the case of low concentrations of dihydroneopterin (see above), and the standard deviations of the recoveries were also satisfying. The manganese dioxide oxidation procedure was also compared with the iodine oxidation procedure according to

Fukushima and Nixon (11). Very similar quantitative results were obtained.

The limit of detection was estimated using biopterin. With the spectrofluorometer PE 650-10 S (at 350/450 nm, slits 10/20, range 1) 10 pg biopterin injected could be observed at a signal to noise of 5:1. A similar sensitivity was observed using the Schoeffel filter fluorometer. The selectivity of both detectors was comparable, when the special emission filter B-40 was used in the Schoeffel instrument. This was controlled by measuring series of urinary samples with both detectors connected in series.

TABLE 6

Reproducibility of analysis of urinary biopterin (B) and neopterin (N) (HPLC variant I) and recovery of tetrahydrobiopterin (BH_4), dihydrobiopterin (BH_2) and dihydroneopterin (NH_2) added to three different urine specimen.

No. of estimations n	Compound added	conc. (µmol/l)	Concentration found	mean (µmol/l)	coefficient of variation (%)	Recovery %
4	-	-	B	0.30	3.7	-
10	BH_4	0.38	B	0.66	2.8	95.1± 5.2
10	BH_4	3.85	B	4.05	2.5	97.4± 2.7
4	-	-	B	0.65	3.6	-
10	BH_2	0.40	B	1.03	2.9	94.2± 7.4
10	BH_2	4.05	B	4.34	1.9	91.2± 2.0
4	-	-	N	0.20	11.2	-
10	NH_2	0.36	N	0.52	7.4	89.1±10.0
10	NH_2	3.49	N	3.44	3.7	93.0± 3.6

HPLC of yellow fluorescing pterins

Separation of the sepia analogs with HPLC variant IV is shown in Fig. 5. Sepiapterin and its analogs are very sensitive

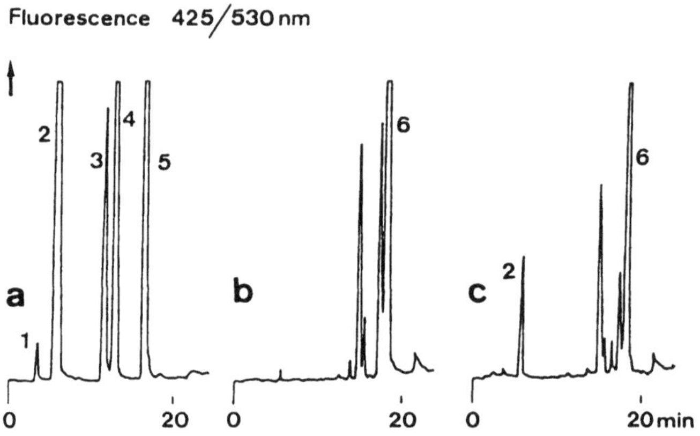

Fig. 5. Reverse phase high-performance liquid chromatograms of yellow fluorescing pterins. a) reference mixture: xanthopterin (1), 3'-hydroxysepiapterin (2), 2'-deoxy-3'-hydroxysepiapterin (3), sepiapterin (4), 2'-deoxysepiapterin (5), riboflavin (6); injected amount: 10 ng of nos. 2,4, and 5. b) urine of a patient with dihydropteridine reductase deficiency. c) urine of a patient with defective dihydrobiopterin biosynthesis.

against light and oxygen. The compounds decomposed markedly, particularly in diluted solutions like the standard solution, when not kept under inert gas and protected from light. Sepiapterin was oxidized to 7,8-*de*hydrosepiapterin (= 6-lactylpterin), a compound which practically did not fluoresce but decomposed in a photoreaction (e.g. within the fluorometer) to pterin-6-carbonic acid. The same was observed with 3'-hydroxysepiapterin. It is therefore essential to analyze only fresh samples or samples kept strictly under inert gas and protected from light. Chromatograms of urines from patients with deficiencies in dihydropteridine reductase and dihydrobiopterin, respectively, are also shown in Fig. 5. Interestingly, practically no xanthopterin was found in the urine of the patient with dihydropteridine reductase deficiency. A rather high concentration of 3'-hydroxysepiapterin was observed in patients with defective dihydrobiopterin biosynthesis (16). Urinary riboflavin could be measured by this procedure as well (Fig. 5b,c). As outlined already, the HPLC column has to be cleaned after a limited number of injections of native urine. Peak broadening and increase of back pressure was observed already after approx. 20 injections.

Yellow fluorescing and blue fluorescing pterins could be measured simultaneously by connecting two fluorimeters in series. In this way it was possible to measure the concentrations of oxidized pterins and of 3'-hydroxysepiapterin in native urine. This technique may be useful in combination with the difference measurement of tetrahydrobiopterin by acidic/alkaline oxidation (11).

HPLC of tetrahydrobiopterin

Chromatograms of synthetic tetrahydrobiopterin preparations are shown in Fig. 6. The 6-S,R enantiomers could be separated easily. The natural isomer was moving faster than the unnatural. Dihydrobiopterin eluted later on but was only hardly detectable at the amperometric setting at 200 mV. Also, the thiols added for stabilization of tetrahydrobiopterin solutions showed a low response at 200 mV and did not interfere. The method was applied for measurement of tetrahydrobiopterin in human milk (17). After defatting and deproteinization, only the natural isomer was found (Fig. 6d). A minor peak just before the expected position of the unnatural isomer was stable against oxidation, unlike the tetrahydrobiopterin enantiomers.

Simultaneous analysis of pterins and creatinine

Hausen *et al.* (15) described the simultaneous determination of urinary neopterin and creatinine by HPLC. Creatinine was monitored at 235 nm by a u.v. detector connected in series

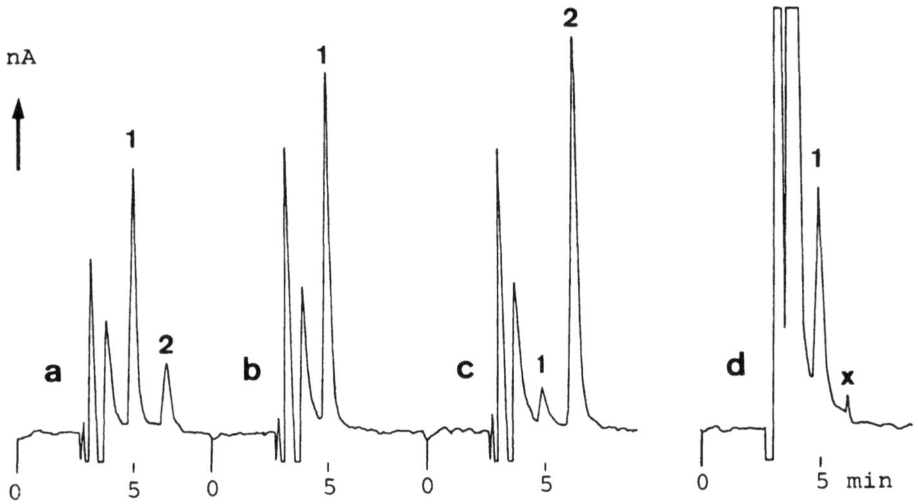

Fig. 6. Reverse phase HPLC separation of 6-S,R enantiomers of synthetic L-erythro-5,6,7,8-tetrahydrobiopterin preparations (a,b,c) and analysis of tetrahydrobiopterin in human milk (d). Natural (1) and unnatural (2) enantiomer; unknown (x). Injected amounts: approx. 2 ng of pure standards; 15 µl human milk.

to the fluorescence detector. This technique can also be applied to HPLC variant I as follows: Solvent A: 7.5 mmol/l potassium phosphate pH 6.3; Solvent B: isopropanol-methanol (1:1, v/v). Creatinine standard: 100 mg/l (0.884 mmol/l) creatinine in 0.1 M HCl (Beckman creatinine standard diluted 1:10 with 0.1 M HCl). The relative fluorescence of pterins decreased by a factor of 1.2-1.3, except that of isoxanthopterin, which decreased by a factor of approx. 2. Pterins moved faster by a factor of approx. 1.4 (compare Table 5, 3% methanol). Isoxanthopterin moved nearer to monapterin. This technique is advantageous in screening for tetrahydrobiopterin deficiency, since all relevant parameters can be measured simultaneously. The creatinine values were not influenced by the oxidation step with manganese dioxide.

Acknowledgements

We are grateful to Prof. M. Viscontini (University of Zurich, Switzerland), Prof. W. Pfleiderer (University of Constance, FRG) and Dr. R.A. Levine (NIH, Bethesda, MD, USA) for gifts of pterins. We thank Miss Ana Matasović and Mr. W. Scherrer for technical assistance. This work was supported by the Swiss National Science Foundation, project no. 3.815-0.79 and 3.850-0.81.

References

1. Bailey, S.W., Ayling, J.E.: High pressure liquid chromatography of substituted pteridines and tetrahydropteridines. In: "Chemistry and Biology of Pteridines." Pfleiderer, W. (ed.). Walter de Gruyter, Berlin. pp 633-643 (1975).

2. Bailey, S.W., Ayling, J.E.: Separation and properties of the 6-diastereoisomers of L-erythro-tetrahydrobiopterin and their reactivities with phenylalanine hydroxylase. J. Biol. Chem. 253, 1598-1605 (1978).

3. Stea, B., Halpern, R.M., Smith, R.A.: Separation of unconjugated pteridines by high-pressure cation-exchange liquid chromatography. J. Chromatogr. 168, 385-393 (1979).

4. Stea, B., Halpern, R.M., Halpern, B.C., Smith, R.A.: Urinary excretion levels of unconjugated pterins in cancer patients and normal individuals. Clin. Chim. Acta. 113, 231-242 (1981).

5. Schlesinger, P., Watson, B.M., Cotton, R.G.H., Danks, D.M.: Urinary dihydroxanthopterin in the diagnosis of malignant hyperphenylalaninemia and phenylketonuria. Clin. Chim. Acta. 92, 187-195 (1979).

6. Fukushima, T., Shiota, T.: Pterins in human urine. J. Biol. Chem. 247, 4549-4556 (1972).

7. Niederwieser, A., Curtius, H.-Ch., Bettoni, O., Bieri, J., Schircks, B., Viscontini, M., Schaub, J.: Atypical phenylketonuria caused by 7,8-dihydrobiopterin synthetase deficiency. Lancet I, 131-133 (1979).

8. Karobath, M., Röthler, F., Heckl, K.: Quantitative determination of unconjugated pterins in biological extracts by gas chromatography-mass fragmentography. In: "Recent Developments in Mass Spectrometry in Biochemistry and Medicine." (Frigerio, A., ed.) Vol. 1, 439-501 (1978).

9. Niederwieser, A., Bieri, J., Blehova, B., Curtius, H.-Ch., Matasović, A., Rey, F., Schircks, B., Viscontini, M.: Urinary pterins in atypical phenylketonuria (Abstract). Pediat. Res. 14, 175 (1980).

10. Niederwieser, A., Curtius, H.-Ch., Gitzelmann, R.,*et al.*: Excretion of pterins in phenylketonuria and phenylketonuria variants. Helv. Paed. Acta. 35, 335-342 (1980).

11. Fukushima, T., Nixon, J.C.: Analysis of reduced forms of biopterin in biological tissues and fluids. Anal. Biochem. 102, 176-188 (1980).

12. Kohashi, M., Tomita, K., Iwai, K.: Analysis of unconjugated pterins in food resources and human urine. Agric. Biol. Chem. 44, 2089-2094 (1980).

13. Rokos, H., Rokos, K., Frisius, H., Kirstaedter, H.J.: Altered urinary excretion of pteridines in neoplastic disease. Determination of biopterin, neopterin, xanthopterin, and pterin. Clin. Chim. Acta 105, 275-286 (1980).

14. Dhondt, J.L., Largilliere, C., Ardouin, P., Farriaux, J.P., Dautrevaux, M.: Diagnosis of variants of hyperphenylalaninemia by the determination of pterins in urine. Clin. Chim. Acta 110, 205-214 (1981).

15. Hausen, A., Fuchs, D., König, K., Wachter, H.: Determination of neopterin in human urine by reversed-phase high-performance liquid chromatography. J. Chromatogr. 227, 61-70 (1982).

16. Niederwieser, A., Matasović, A., Curtius, H.-Ch., Endres, W., Schaub, J.: 3'-Hydroxysepiapterin in patients with dihydrobiopterin deficiency. FEBS Lett. 118, 299-302 (1980).

17. Niederwieser, A., *et al.* (in preparation).

HPLC OF NEOPTERIN PHOSPHATES

Application to the Study of GTP-Cyclohydrolases

N. Blau, R. Bosshard, U. Redweik and A. Niederwieser

Division of Clinical Chemistry, Dept of Pediatrics, University of Zurich, Steinwiesstr. 75, CH-8032 Zurich, Switzerland

Introduction

GTP-Cyclohydrolases are enzymes that catalyze the removal of C-8 from guanosine triphosphate (GTP) to give a phosphorylated 2,5-diamino-4-oxo-6-(ribosylamino)pyrimidine derivative and formic acid. This reaction is the first step in the biosynthesis of a number of important cofactors, namely, of tetrahydrofolic acid (1,2), tetrahydrobiopterin (Ref. 2, see also Chapter 1 p27 in this book) and riboflavin (1). Also queuine, a modified base (a deazaguanine) incorporated posttranscriptionally into eucaryotic transfer RNA (3) seems to be formed by a GTP-cyclohydrolase (4). The structure of the pyrimidine derivative has not yet been clarified unequivocally in every case; it may not be the same from different GTP-cyclohydrolases and the intermediate can be transformed into a further metabolite, possibly by the same enzyme or even nonenzymatically.

In the reaction of GTP-cyclohydrolase I (EC 3.5.4.16) (5) also referred to as dihydroneopterin triphosphate synthetase (6), the pyrimidine derivative cannot be isolated and is directly transformed into D-erythro-7,8-dihydroneopterin triphosphate (NH_2TP) over an Amadori rearrangement of the amino sugar (7) and ring closure to the pyrazine moiety of the pterin. NH_2TP is employed in the biosynthesis of tetrahydrofolic acid and tetrahydrobiopterin. It is an unstable compound which, in the pres-

Biochemical and Clinical Aspects of Pteridines · Volume 1
Cancer · Immunology · Metabolic Diseases

ence of oxygen, decomposes rapidly into numerous products.

GTP-cyclohydrolase II from E.coli produces 2,5-diamino-4-oxo-6-(ribosylamino)pyrimidine 5'-phosphate (PydP), pyrophosphate and formic acid from GTP (8). In rat brain, a GTP-cyclohydrolase A-I and a magnesium-dependent A-II have been described (9), which hydrolyzed GTP to 2-amino-6-(5'-triphosphoribosyl)amino-5- or6-formamido-4-oxopyrimidine which was transformed into NH_2TP by a further enzyme, named again D-erythro-7,8-dihydroneopterin triphosphate synthetase (enzyme B), a basic protein with known amino acid sequence (10).

The commonly used method for the measurement of GTP-cyclohydrolase activity is based on the release of radioactive formic acid from [^{14}C]8-GTP (11) and gives no information about the nature of the other products. These - usually unstable - compounds have been isolated using time consuming column chromatographic procedures. We thought it might be useful to employ a faster and high resolution chromatographic technique to measure the primary products more specifically. Undoubtedly, this will be necessary to clarify some of the discrepancies in literature of this field.

Although there is a lot of information on separation of nucleotides by high-performance liquid chromatography (HPLC), this powerful technique has not yet been used for the separation of the primary products of GTP-cyclohydrolases, namely, the pyrimidine derivatives mentioned above and the neopterin phosphates.

We have developed an ion-pair reverse phase HPLC of neopterin phosphates for an assay of GTP-cyclohydrolase I activity in human and rat liver (12). In the following, we describe the application of this technique for the measurement of GTP-cyclohydrolases I and II in mammalian and bacterial systems.

Materials and Methods

Chemicals

Guanine, guanosine, guanosine 5'-monophosphate (GMP), guanosine 5'-diphosphate (GDP), guanosine 5'-triphosphate (GTP), triethylamine, N,N'-dicyclohexyl carbodiimide and 4-hydroxy-2,5,6-triamino pyrimidine sulfate were purchased from Fluka AG (Buchs, Switzerland); alkaline phosphatase from Boehringer GmbH (Mannheim, FRG); pterins from Laboratory Dr. B. Schircks (Wettswil a.A., Switzerland); Sephadex G-25 and DEAE-Sephadex A-50 from Pharmacia (Uppsala, Sweden); Ultrogel AcA-34 from LKB (Bromma, Sweden); $[^{14}C]$8-GTP from Amersham (Amersham, UK); frozen ½ log cells of Escherichia coli B and LiChrosorb RP-8, 5 μm, from Merck (Darmstadt, FRG).

GTP-Cyclohydrolase I and II from E.coli

The enzymes were partially purified using modifications of the procedures of Yim and Brown (5) and Foor and Brown (8), respectively. The frozen E.coli cells (300 g, moist weight) were mixed with 380 ml of 0.3 M Tris-HCl buffer (pH 8.0) and allowed to thaw at room temperature. After standing at $4^{o}C$ for 3 days, the mixture was incubated at $31^{o}C$ for 20 min with 300 mg of lysozyme and 33 mg of DNase A, cooled to $4^{o}C$ and centrifuged at 10800 x g for 75 min. The supernatant was saturated to 35% with ammonium sulfate, stirred for 5 min and centrifuged at 10800 x g for 40 min. The resulting supernatant was then saturated to 50% with ammonium sulfate and after 15 min centrifuged at 10800 x g for 45 min. The precipitate was suspended in 25 ml of 0.05 M Tris-HCl buffer (pH 8.0) containing 0.3 M KCl and dialyzed overnight against the same buffer. After centrifugation the enzyme solution was applied to a column (81x5 cm) of Sephadex G-75 equilibrated with 0.05 M Tris-HCl buffer (pH 8.0) containing 0.3 M KCl. The proteins were eluted with the same buffer at a flow rate of 0.75 ml/min and 15.2 ml fractions were collected. The fractions were analyzed for GTP-cyclohydrolase activity by measuring the release of formic acid from GTP (see below). Fractions 41-58 with GTP-cyclohydrolase I activity (Fig. 1) were combined and concentrated to 25 ml by ultrafiltration.

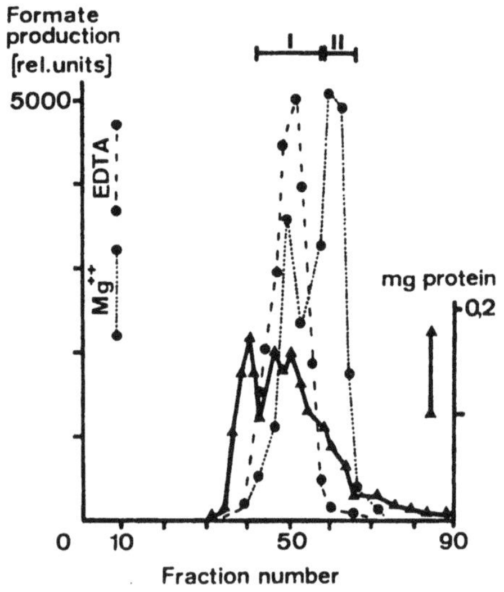

Fig. 1. Gel filtration of GTP-cyclohydrolase I and II in dialyzed extract of E.coli on Sephadex G-75. Details see text. Protein is given in mg/fraction.

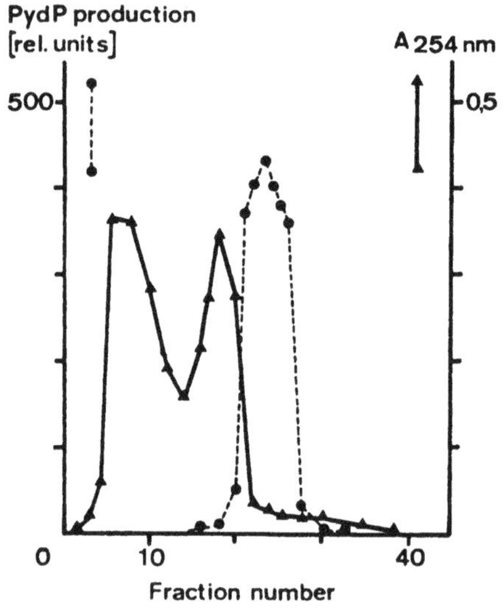

Fig. 2. Chromatography of GTP-cyclohydrolase II from fractions no.59-66 (Sephadex G-75, Fig. 1) on DEAE Sephadex A-50. Enzyme activity was measured by HPLC of pyrimidine derivatives (PydP) formed. Protein was measured by absorbance at 254 nm.

This solution was applied to a column (50x5 cm) of Ultrogel AcA-34 previously equilibrated with 0.05 M Tris-HCl buffer (pH 8.0) containing 0.3 M KCl. The proteins were eluted with the same buffer at a flow rate of 0.75 ml/min and fractions of 15 ml were collected. Fractions 47-63 containing GTP-cyclohydrolase I activity were reduced to a volume of 15 ml. This preparation had a specific activity of 46 units per mg of protein and was used for preparation of NH_2TP.

For purification of GTP-cyclohydrolase II, fractions 59-66 from Sephadex G-75 gel filtration (Fig. 1) were combined and applied onto a 31x2.5 cm column of DEAE Sephadex A-50 equilibrated with 0.025 M Tris-HCl buffer (pH 8.0) containing 0.25 M NaCl. The flow rate was 0,5 ml/min and the first 10 fractions (210 ml) were eluted with the same buffer. From fraction 11 GTP-cyclohydrolase II was eluted with the same buffer containing 0.35 M NaCl. Fractions 21-26 containing GTP-cyclohydrolase II activity (Fig. 2) were combined and reduced to a volume of 15 ml. This preparation contained partially purified GTP-cyclohydrolase II and was used for the preparation of ribitylpyrimidines.

Neopterin phosphates

NH_2TP was prepared from GTP using partially purified GTP-cyclohydrolase I from E.coli (see above). The incubation mixture contained GTP (1.35 µmol), Tris-HCl buffer pH 8.0 (42 µmol), KCl (0.2 mol) and enzyme, in a total volume of 0.75 ml. Incubation was at 37°C for 3 hours in the dark. Neopterin 3'-triphosphate (NTP) was prepared from NH_2TP by iodine oxidation: 500 µl of a solution containing NH_2TP was mixed with 50 µl of acidic iodine solution (1% I_2, 2% KI, 1 M HCl). After standing for 30 min at room temperature, the mixture was centrifuged and 50 µl of 1% ascorbic acid solution was added to remove excess iodine. The mixture was evaporated to dryness. The residue was dissolved in 100 µl of mobile phase (see below) and injected for HPLC (see below). The NTP containing fraction was collected and stored at -20°C.

Neopterin 3'-monophosphate (NMP) and neopterin 3'-diphosphate (NDP) were isolated by HPLC after partial hydrolysis of NTP: 500 µl of the NTP solution was left at room temperature overnight, reduced to a volume of 100 µl in vacuo and separated by HPLC. NMP and NDP containing fractions were collected and stored at -20°C. The neopterin phosphates were identified as neopterin after hydrolysis with alkaline phosphatase: Fractions were mixed with an equal volume of 0.2 M Tris-HCl buffer (pH 9.8) and 3 units of alkaline phosphatase and incubated for 20 min at 37°C. 10 µl of the mixture was injected for measurement of neopterin by HPLC.

HPLC

The HPLC system consisted of a Model 110 pump and Model 420 programmer (Altex, Berkeley, CA, USA), WISP 710A automatic sampler (Waters Assoc., Milford, MA, USA), Model LC55 spectrophotometer (Perkin Elmer, Oak Brook, IL, USA) for detection of nucleotides at 254 nm and 287 nm, FS 950 filter fluorometer (Kratos, Westwood, NJ, USA) for detection of pterins at 365/446 nm using Kratos FSA 110 excitation lamp and B-40 emission filter with 10 nm bandpass (Balzers, Principality of Liechtenstein), Model 3390A integrator (Hewlett Packard, Avondale, USA) and W+W Model 314 recorder (Kontron, Zurich, Switzerland).

HPLC was performed on LiChrosorb RP-8, 5 µm, in a precolumn (4.6x40 mm) connected in series with an analytical column (4.6x250 mm) at room temperature and a flow rate of 1 ml/min. The mobile phase was a triethylammonium phosphate buffer (pH 7.0) of the following composition: Isopropanol-85%-phosphoric acid-triethylamine-water (3:3:10:984, v/v). Peaks are identified by comparing retention times with the reference compounds or absorbance ratios at 254 and 287 nm. Compounds are measured by peak integration using external standards.

Assay of GTP-cyclohydrolase I and II activity: Release of formic acid

The method is based on production of radioactive formic acid according to Guroff and Strenkoski (11) modified. GTP-cyclohydrolase I activity was measured in the presence of EDTA and

GTP-cyclohydrolase II (+ I) activity in the presence of Mg^{++}.

The incubation mixture contained enzyme, 1.8 nmol [^{14}C]8-GTP (0.1 µCi), 40 nmol carrier GTP, 0.8 µmol EDTA (GTP-cyclohydrolase I) or $MgCl_2$ (GTP-cyclohydrolase II), approx. 10 µmol KCl and 10 µmol Tris-HCl buffer (pH 7.6) in a total volume of 0.35 ml. The incubation was carried out in the dark at 37°C for 1 hour. The reaction was stopped by addition of 2 ml of 0.1 M HCl and the formic acid released was extracted with 3 ml of ethyl acetate. One ml of the ethyl acetate was mixed with 10 ml of Lumagel, for radioactivity measurement (cpm should not exceed 10,000). One unit corresponds to 1 nmol of the radioactive formic acid released from [^{14}C]8-GTP per hour. The enzyme activity was calculated using the following formula:

$$U/ml = \frac{cpm}{V} \cdot F$$

whereby

$$F = \frac{3 \cdot D}{Y \cdot R \cdot spec.act. \cdot 2220} \approx 1.6$$

and V=Volume of enzyme solution in µl; D=Dilution factor of [^{14}C]8-GTP with carrier GTP (≈23); Y=Yield of formic acid extracted into ethyl acetate (=0.4), and R=Counting recovery (in our instrument, cpm=0.89 dpm).

Assay of GTP-cyclohydrolase I activity: HPLC of neopterin-triphosphate

The incubation conditions were according to Viveros *et al.* (13). The NH_2TP formed was oxidized to NTP and measured by HPLC. The incubation mixture contained GTP (2 mM), EDTA (2.5 mM), KCl (0.3 M), glycerol (10%), Tris-HCl buffer pH 7.8 (0.1 M) and enzyme in a total volume of 0.45 ml. The mixture was incubated for 90 min at 37°C in the dark. The reaction was stopped by addition of 50 µl of a 1 M HCl solution containing 1% iodine and 2% KI. After standing at room temperature for 30 min, the sample was centrifuged and 50 µl of 1% ascorbic acid were added to the clear supernatant. The pH was adjusted to 7.0 using 1 M NaOH and the sample was ready for HPLC measurement of NTP. Measurement of GTP-cyclohydrolase activity in liver required purification of the homogenate on a small Sephadex G-25 column.

Assay of GTP-cyclohydrolase II activity: HPLC of pyrimidine derivatives

The incubation mixture was the same as for GTP-cyclohydrolase I activity measurement (HPLC method), except that $MgCl_2$ (5 mM) was used instead of EDTA. However, after incubation, 10 µl of the mixture was directly applied for HPLC, without oxidation.

Results and Discussion

A mixture of standard guanine nucleotides (Fig. 3A) and neopterin phosphates (Fig. 3B) could be separated in less than 20 min using 44 mM triethylammonium phosphate buffer pH 7.0 containing 0.3% isopropanol as the mobile phase. Use of triethylammonium

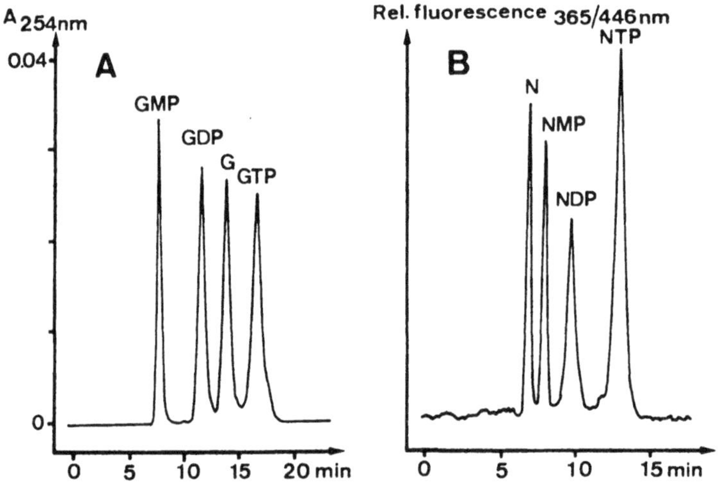

Fig. 3. Chromatographic profile of guanosine and guanosine phosphates (A) and neopterin and neopterin phosphates (B).

phosphate yielded good results, comparable with the commonly used tetrabutylammonium phosphate which is expensive. Elution is accelerated by addition of isopropanol to the mobile phase and only a slight change in its concentration results in marked differences in retention times (Fig. 4). Pyrimidine phosphate

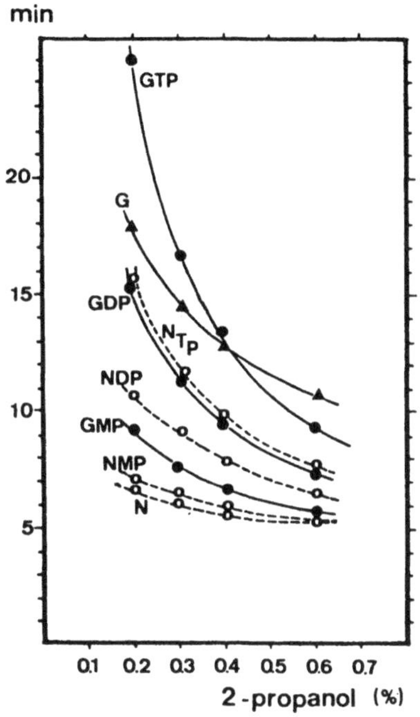

Fig. 4 Effect of isopropanol concentration in the mobile phase on the retention times of neopterin phosphates and guanosine phosphates.

derivatives with very short elution times could be retained by omitting the organic solvent in the mobile phase. GTP-cyclohydrolase I activity could be measured in liver with as little as 15 mg of tissue. Therefore, the method can be used to measure the enzyme activity in liver biopsies. The application of the method on human liver homogenate is demonstrated in fig. 5, showing chromatograms of enzyme blank (mixture incubated without substrate), substrate blank (mixture incubated without enzyme) and complete assay mixture. After incubation with GTP, no other fluorescent peaks except NTP could be detected. NTP could be partially hydrolyzed to NDP, NMP and neopterin, by

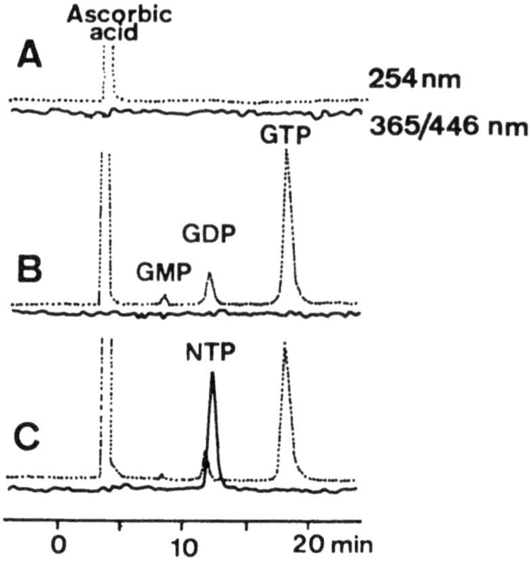

Fig. 5. GTP-cyclohydrolase I assay in human liver by HPLC: Enzyme blank (A), substrate blank (B) and assay mixture (C), after incubation and oxidation with iodine.

incubation with alkaline phosphatase at room temperature for 20 min. This could be controlled by HPLC (Fig. 6).

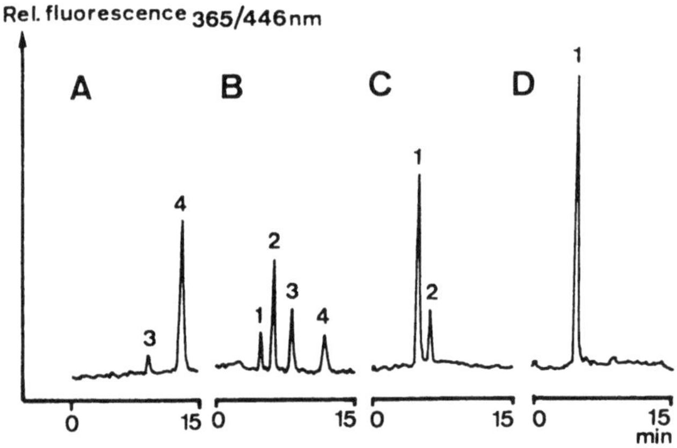

Fig. 6. Hydrolysis of NTP with alkaline phosphatase at room temperature as a function of time. Incubation time: 3 min (A), 20 min (B), 45 min (C) and 120 min (D). Neopterin (1), NMP (2), NDP (3), NTP (4).

Total hydrolysis of NTP to neopterin was achieved by treatment with alkaline phosphatase for 20 min at 37°C.

Incubation of GTP in the presence of EDTA with partially purified GTP-cyclohydrolase I from human liver (after 30-50% ammonium sulfate precipitation, chromatography on DEAE-Sephacel and Sephacryl S-300), gave only NH_2TP, and no other products like pyrimidine phosphate derivatives could be detected. This is contrary to the findings of Gal *et al.* (9), who postulated the existence of two enzymes converting GTP to 2-amino-6-(5'-triphosphoribosyl)amino-5- or 6-formamido-4-oxopyrimidine ($FPydP_3$) in rat brain. We found no indication for the production of $FPydP_3$ in human liver homogenate.

Combination of two u.v. detectors (254 and 287 nm) and a fluorometer can be useful for monitoring the substrate and reaction products at any step of the reaction. For instance, in a trial for measuring GTP-cyclohydrolase II activity in human liver homogenate, we found that under the conditions employed in the assay (presence of Mg^{++}), GTP was rapidly converted into uric acid. The uric acid was identified by enzymatic test with uricase, u.v.-spectrum, HPLC and GC-MS. Interestingly, this compound has an absorption maximum at 292 nm, very similar to pyrimidine derivatives! Investigations with partially purified human liver homogenate (30-50% ammonium sulfate followed by DEAE-Sephacel chromatography) have shown that the uric acid producing activity was within a narrow protein fraction.

According to Foor and Brown (8), GTP-cyclohydrolase II from E.coli produces PydP from GTP. We were surprised to find two different products, PydP I and PydP II, when partially purified GTP-cyclohydrolase II from E.coli was incubated with GTP (Fig. 7). Using [α-^{32}P]GTP, both products were shown to be phosphorylated. Treatment with alkaline phosphatase changed their retention times and removed the [^{32}P] label. Oxidation

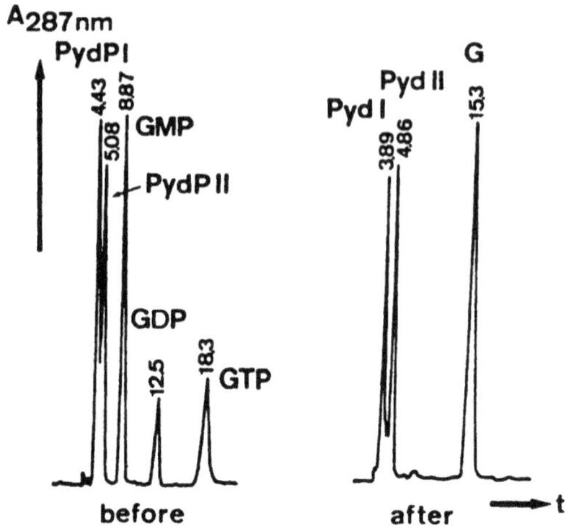

Fig. 7. GTP-cyclohydrolase II (E.coli) assay by HPLC, before and after hydrolysis with alkaline phosphatase. Production of two phosphorylated pyrimidine derivatives (PydP I and PydP II). Retention times in minutes.

with meta-periodate gave pterin, and treatment with 2,3-butanedione gave 6,7-dimethylpterin. Although the exact structure of these compounds could not yet be clarified, it seems that both compounds are isomers.

Acknowledgements

We thank Mr. W. Leimbacher and Mr. W. Staudenmann for skillful technical assistance. This work was supported by the Swiss National Science Foundation, project no. 3.815-0.79 and 3.933-0.78.

References

1. Brown, G.M., Williamson, J.M.: Biosynthesis of folic acid, thiamine, and pantothenic acid. Adv. Enzymol. 53, 345-381 (1982).

2. Blakeley, R.L. (ed.): "The Biochemistry of Folic Acid and Related Pteridines." Northe Holland Research Monographs Frontiers of Biology Vol. 13, North Holland Publ. Co., Amsterdam, London (1969).

3. Katze, J.R., Basile, B., McCloskey, J.A.: Queuine a modified base incorporated posttranscriptionally into eucaryotic transfer RNA: Wide distribution in nature. Science 216, 55-216 (1982).

4. Jacobson, K.B., Farkas, W.R., Katze, J.R.: Presence of queuine in *Drosophila melanogaster*: Correlation of free pool with quenosine content of tRNA and effect of mutations in pteridine metabolism. Nucleic Acids Res. 9, 2351-2366 (1981).

5. Yim, J.J., Brown, G.M.: Characteristics of guanosine triphosphate cyclohydrolase I purified from *Escherichia coli*. J. Biol. Chem. 251, 5087-5094 (1976).

6. Jackson, R.J., Shiota, T.: Identification of the isomer of dihydroneopterin phosphate synthesized by two enzymes fractions from *Lactobacillum plantarum*. J. Biol. Chem. 246, 7454-7459 (1971).

7. Wolf, W.A., Brown, G.M.: The biosynthesis of folic acid X. Evidence for an Amadori rearrangement in the enzymatic formation of dihydroneopterin triphosphate from GTP. Biochim. Biophys. Acta 192, 468-478 (1969).

8. Foor, F., Brown, G.M.: GTP-cyclohydrolase II from *Escherichia coli* . Meth. Enzymol. 66, 303-307 (1980).

9. Gal, E.M., Nelson, J.M., Sherman, A.D.: Biopterin III. Purification and characterization of enzymes involved in the cerebral synthesis of 7,8-dihydrobiopterin. Neurochem. Res. 3, 69-88 (1978).

10. Gal, E.M., Dawson, M.R., Dudley, D.T., Sherman, A.D.: Biopterin VI. Purification and primary amino acid sequence of mammalian D-erythro-7,8-dihydroneopterin triphosphate synthetase. Neurochem. Res. 4, 605-626 (1979).

11. Guroff, G., Strenkoski, C.A.: Biosynthesis of pteridines and of phenylalanine hydroxylase cofactor in cell-free

extracts of *Pseudomonas*. J. Biol. Chem. 241, 2220-2227 (1966).

12. Blau, N., Niederwieser, A.: GTP-cyclohydrolase I assay in human and rat liver using high-performance liquid chromatography of neopterin phosphates and guanine nucleotides. Anal. Biochem. (in press).

13. Viveros, O.H., Lee, C.L., Abou-Donia, M.M., Nixon, J.C., Nichol, C.A.: Biopterin cofactor biosynthesis: Independent regulation of GTP cyclohydrolase in adrenal medulla and cortex. Science 213, 349-350 (1981).

PTERIDINES AS TUMOUR MARKERS? ATTEMPT OF EVALUATION BY HPLC AND A RADIOIMMUNOASSAY FOR NEOPTERIN

Katharina Rokos and Hartmut Rokos

Research Laboratory, Henning Berlin GmbH,
Komturstraße 19-20, D-1000 Berlin 42, Germany

Hildegard Frisius

Department of Internal Medicine, Martin-Luther-Krankenhaus,
Berlin, Germany

Introduction

In recent years there has been a growing number of reports on altered pteridine metabolism in a variety of diseases other than atypical PKU, where deficiencies in pteridine synthesis have been already well established, the most significant disturbance being an elevation of neopterin levels (1-14). Most information is on patients with neoplastic disease, but some data on patients with renal insufficiencies, viral disease, neurological disease and others are available, too. Data so far obtained are difficult to compare since methods for measuring pteridines vary widely. Crithidia fasciculata assay and phenylalanine hydroxylase assay, which were the first assays used are sensitive and applicable to body fluids with low biopterin content as serum and CSF but measure biopterin only. The HPLC methods have the advantage of determining simultaneously all pteridines present, but so far satisfactory results were obtained mostly in urine with high pteridine concentrations. A radioimmunoassay for biopterin was recently developed and first results for biopterin levels in several diseases reported (15, 16). This method is very sensitive, specific and suitable for rapid evaluation of large numbers of samples. We

Biochemical and Clinical Aspects of Pteridines · Volume 1
Cancer · Immunology · Metabolic Diseases

report preliminary results with a newly developed radioimmunoassay for neopterin in serum and urine together with HPLC data on excreted pteridines.
The aim of this study is to evaluate whether determination of pteridine levels, especially neopterin, might be useful in cancer detection and in evaluation of cancer therapy with special regard to other conditions which might elevate pteridine levels and so interfere with interpretation of results.

Materials and Methods

Urine from patients and controls was mostly obtained in the morning. Urine and serum samples were either used immediately or stored at -20 °C. Pteridines in urine and serum were stable for at least several weeks at -20 °C. The oxidizable amounts of pteridines decreased slightly during 24 h at room temperature in the dark. All urinary pteridine concentrations are given relative to urinary creatinine concentrations, determined colorimetrically (Merckotest Creatinine-Kit, Merck).
Oxidation of pteridines was done according to (17) using acidic iodine with various oxidation times. Recovery of pteridines added to urine and serum was > 90 %. Neopterin and Biopterin values were well reproducible.
High pressure liquid chromatography (HPLC). Pteridine concentrations were measured partly according to a HPLC procedure previously described (7) - method A -, or according to (17, 18) with a Knauer Model FR-30 pump and a Farrand MK 2 fluorescence detector. Urine was injected without further prepurification - method B.
Radioimmunoassay (RIA). 5 µl of urine (or neopterin standards) are incubated simultaneously with I-125-neopterin tracer (ca. 30000 cpm), first (anti-neopterin, rabbit) and second (goat anti-rabbit-γ-globulin) antibody for 2 h at 37 °C. The precipitate is separated by centrifugation, washed, and then counted in a γ-counter (Berthold BF 6000).

	ng Neopterin/ mg Creatinine mean ± SD (range)	ng Biopterin/ mg Creatinine mean ± SD (range)	N/B ratio mean ± SD (range)
Healthy Males	1045 ± 345	1310 ± 535	0.9 ± 0.33
n = 16 (20 - 63)	(515 - 1860)	(480 - 2400)	(0.5 - 1.5)
Healthy Females	1335 ± 427	1600 ± 642	0.92 ± 0.32
n = 23 (24 - 70)	(680 - 2070)	(766 - 3980)	(0.25 - 1.8)
Healthy Children	1328 ± 452	2096 ± 623	0.67 ± 0.28
n = 9 (6 - 15)	(790 - 2250)	(1220 - 3230)	(0.4 - 1.15)

Table 1. Total Urinary Neopterin and Biopterin in Healthy Controls

Male n = 17	1462 ± 275	1547 ± 479	1.04 ± 0.27
	(1000 - 1860)	(810 - 2300)	(0.65 - 1.3)
Female n = 20	1355 ± 326	2342 ± 598	0.62 ± 0.22
	(800 - 1880)	(1550 - 3980)	(0.25 - 1.1)

Table 2. Variation of Pteridine Excretion in two Healthy Controls (from June to December 1981)

Serum neopterin is determined analogously using 50 µl serum (or neopterin standards), incubation overnight at 37°C.

Results

Urinary pteridines of healthy controls after acidic iodine oxidation are shown in Table 1. These normal values for neopterin are higher than those reported by some authors (19,20), but lower than those in (9). Neopterin values measured without oxidation (21) correlated well with those obtained after ion exchange chromatography (7) but not with (6,11). For comparison the different state of oxidation of neopterin has to be taken into account, since some authors report only direct and some only total values. Total biopterin values are in good agreement with other authors (19,20).
Pteridine excretion varies widely in each individual (Table 2). Especially in females care has to be taken (10), since values might be elevated just before or during menstruation (Figure 1). In cycle 1 ovulation occured (as confirmed by temperature ele-

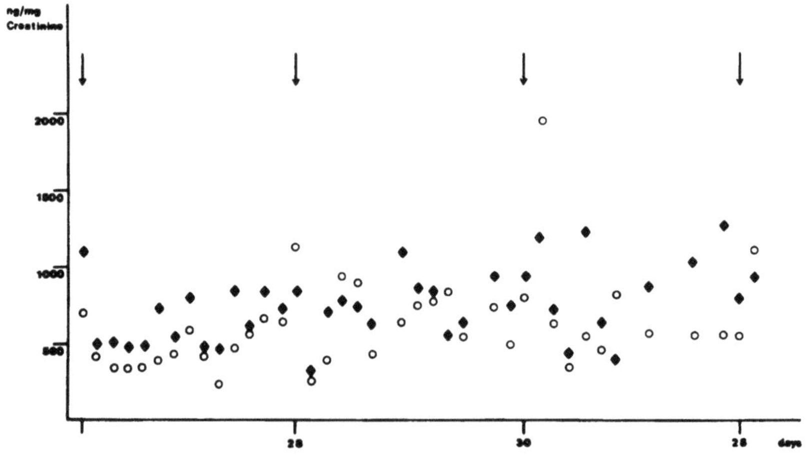

Fig. 1. Urinary pteridines (o neopterin, ◆biopterin) in 3 consecutive menstrual cycles.

	ng Neopterin/ mg Creatinine mean ± SD (range)	ng Biopterin/ mg Creatinine mean ± SD (range)	N/B ratio mean ± SD (range)	n positive* (n total)	χ^2 (P)
Leukemia, Lymphoma	6082 ± 5706 (1670 - 22650)	1314 ± 841 (260 - 2560)	6.2 ± 5.3 (0.8 - 21.1)	13 (15)	< 0.001
Breast Cancer	1594 ± 577 (1050 - 2650)	1477 ± 1124 (470 - 3733)	1.94 ± 1.8 (0.3 - 4.8)	2 (7)	NS
Gastrointestinal Cancer**	2924 ± 1728 (870 - 5340)	1922 ± 835 (1180 - 4140)	1.69 ± 1.0 (0.35 - 3.5)	5 (10)	< 0.02
Various***	2669 ± 1371 (930 - 5700)	1530 ± 850 (670 - 3390)	2.06 ± 1.7 (0.45 - 4.8)	6 (10)	< 0.002

* positive Neopterin > 2000 ng/mg Creatinine (males), > 2400 ng/mg Creatinine (females) and/or Neopterin/Biopterin ratio > 2.0

** Sigma (3), Rectum (3), Gastric (2), Pancreas (3)

*** Ovary (3), Cervix, Bronchus (2), Thyroid (2), Urinary Bladder, Prostate

Table 3. Total Urinary Neopterin and Biopterin in Cancer Patients

Other patients n = 32 (16 - 90)	1952 ± 1440 (660 - 7880)	1630 ± 700 (385 - 3620)	1.75 ± 3.5 (0.4 - 20.5)
Renal diseases n = 6	2731 ± 1377 (490 - 4300)	645 ± 187 (440 - 920)	4.9 ± 3.6 (0.8 - 10.7)

Table 4. Total Urinary Neopterin and Biopterin in patients with non-malignant diseases

vation); during cycle 2 and 3 a slight common cold caused higher fluctuation of pteridine values (method A, upper limit of normal range 1000 ng neopterin/mg creatinine). In four other females results were less conclusive. Similar results for serum biopterin values were reported (8).
Currently, we regard total neopterin values above 2000 ng/mg creatinine for males and 2400 ng/mg creatinine for females and/or a total neopterin/total biopterin ratio above 2.0 as abnormal. Total pteridine values seem to us more reliable than direct values, since they are better reproducible if samples are not processed immediately.
In table 3 pteridine excretion of cancer patients is shown. In some types of cancers, e.g. leukaemia and lymphoma, neopterin is significantly elevated, whereas in others, especially breast cancer (all stage III and IV) elevation is statistically not significant. In breast cancer, raised neopterin excretion is mostly connected with liver metastases or high serum creatinine, indicating renal insufficiency. These two conditions were previously shown to have significantly raised neopterin excretion (7). Biopterin levels are usually in the normal range, high biopterin values are found in very few cases (7). Very low biopterin values together with a high neopterin/biopterin ratio indicate a grave prognosis.
Table 4 shows results of patients with various other diseases from one of the hospitals. Neopterin levels are elevated as compared with normals, but lower than those of cancer patients. These patients together with healthy adults served as control group for calculation of statistical significance of elevated neopterin excretion of cancer patients. Patients with common cold were laboratory staff and only moderatly affected. Patients with renal disease were excluded from the control group as were patients with proven viral illness. Elevated neopterin excretion in patients with herpes zoster (10), measles and rubella (6,10) were already reported. The elevation of neopterin and N/B during such an illness is shown in Table 5 for common cold and measles (method A).

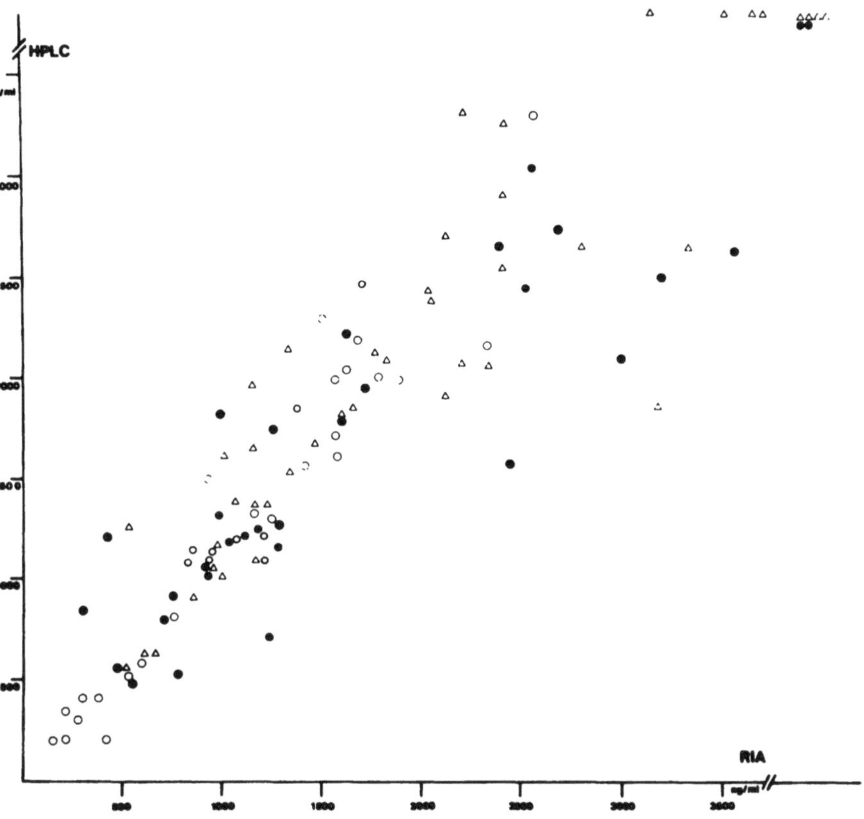

Fig. 2. Urinary neopterin, correlation of HPLC and RIA data
(o controls, △ cancer patients, ● other diseases)

Figure 2 shows the correlation between HPLC and RIA data of urinary neopterin. In most cases correlation is satisfactory; patients who showed gross deviation are under further investigation, if available.

Serum neopterin levels were determined with the radioimmunoassay (Table 6). Normal values are in agreement with those obtained by HPLC (17). The preliminary upper limit of normal neopterin serum levels is set at 4 ng/ml; however, levels in controls vary widely as in urine. Cancer patients had significantly higher serum neopterin, but in other hospital patients

	ng Neopterin/ mg Creatinine	ng Biopterin/ mg Creatinine	N/B
Measles (3 f)			
Basal value	1094	1578	0.7
6th day	6106	687	8.9
8th day	6140	1023	6.0
22nd day	2379	1459	1.6
43rd day	2477	1027	2.4
Bronchitis (38 m)			
Basal value	220	163	1.3
2nd day (no medication)	377	202	1.9
3rd day (Penicillin)	344	258	1.3
5th day (Cotrim)	326	261	1.3
10th day (Cotrim)	287	233	1.2
Common cold (38 m)			
Basal value	468	336	1.4
1st day	2036	483	4.0
5th day	992	477	2.0
9th day	455	310	1.4

Table 5. Pteridine excretion during various common illnesses (method A)

	n	mean	SD	range
Healthy controls	16	2.00	0.68	1.3 - 3.3
Cancer patients	27	7.05	10.5	1.7 - 57.6
Other patients	24	4.06	2.44	1.7 - 10.9
Renal diseases	6	14.8	11.9	3.5 - 70.0

Table 6. Serum Neopterin Levels (ng/ml)

neopterin was elevated too as compared with controls. The highest serum neopterin level (70 ng/ml) so far recorded was in a patient with severe renal insufficiency.

In Figure 3 a correlation of neopterin levels in urine and serum of the same persons was attempted. A fairly close relation was observed in most but not all cases.

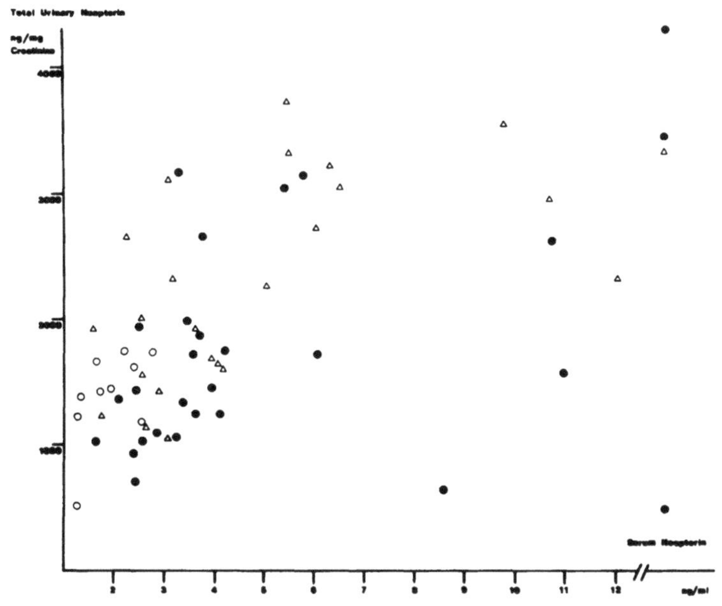

Fig. 3. Correlation of serum and urinary neopterin (o controls, Δ cancer patients, ● other diseases)

Preliminary results (22) on serum neopterin in haematological tumours showed diagnostic potential only in lymphomas (9 out of 12 patients positive with active disease, 0/4 with inactive disease), whereas in active plasmocytomas only 2/9, and in active Hodgkin's disease only 1/4 patients had elevated neopterin levels.

For evaluation of the possible usefulness of neopterin as a tumour marker it was compared with CEA (Table 7). In 96 serum samples CEA levels were elevated in 73 samples (76 %) as compared with neopterin in 25 samples (26 %). Neopterin was high in 5 samples (5 %) with normal CEA values; no correlation of elevated neopterin and CEA levels was found, as anticipated.

In all urine samples investigated HPLC determinations showed a monapterin (threo-neopterin; neopterin = erythro-neopterin) content of 10-15 % of neopterin, as already reported in (7).

Urinary tract infection did not raise the monapterin/neopterin ratio.

	n	CEA > 3 ng/ml	Neopterin > 4 ng/ml
Mamma ca	63	45	15
Colon ca	21	21	8
Various ca	12	7	2
Total	96	73	25

Table 7. Comparison of Elevated CEA and Neopterin Levels in Serum of Cancer Patients

Discussion

The purpose of the investigations on pteridines in urine and serum of cancer patients so far conducted was to determine whether a pteridine was elevated significantly in malignant disease and which pteridine was primarily concerned. Results from various authors have shown, that neopterin is significantly elevated in urine of many but not all cancer patients (5-7, 9-11) as is xanthopterin (7, 9), whereas biopterin elevation is found in few cases (7-10). Since xanthopterin (occuring in urine as 7,8-dihydroxanthopterin) is a degradation product of NH_2, BH_2 and BH_4, and little is known of its biological significance, it seemed logical to concentrate currently on neopterin as a possible tumour marker. Biopterin determinations seem to have their main importance in neurological illnesses (8, 12-14).
HPLC data on neopterin excretion of cancer patients are already available for a limited number of patients (6, 7, 9-11) but serum levels were not reported so far. Data on patients with diseases other than cancer and PKU are scarce (6, 7, 10). With perfection of HPLC methods and development of radioimmuno-

assays for biopterin and neopterin rapid screening of large numbers of samples - urine as well as serum and CSF - is now possible.
Preliminary results with new RIA for neopterin indicated good correlation with HPLC in urine, remaining discrepancies are under investigation. The serum values of neopterin correlated in most but not all cases with urine levels. RIA serum and urine measurements confirmed that neopterin values are frequently raised only in patients with some types of tumours, especially leukemia and lymphoma (7, 10, 11), which are mostly easy to diagnose without tumour markers. Other cancers as breast cancer and cancers of the gastrointestinal tract have much less frequently elevated neopterin levels. A comparison of neopterin with the most widely established tumour marker CEA, which is mainly used in the follow-up of patients with breast cancer and colo-rectal cancer gave disappointing results (Table 7). However, in this set of patients high CEA values are particularly frequent; in a differently composed group of cancer patients where neopterin values are more frequently raised, e.g. pancreatic cancer, liver cancer, hypernephromas (7, 10), comparison with established tumour markers may be more favourable for neopterin. As for tumour markers already in clinical application (e.g. CEA, AFP, ßHCG) the usefulness of neopterin may be in monitoring of progression or regression of disease and not in initial screening for cancer.
The pitfall frequent in determination of tumour markers are false positive results. So search for an exclusion of "false positive" neopterin values was particularly important. Elevated neopterin excretion was already reported in several conditions, e.g. viral disease (6, 10), renal disease (7, 10), haemolytic anaemia (7), severe psoriasis and during menstruation (10). Elevated serum concentrations in haemolytic anaemia and renal disease agreed with these results. The anticipated elevation of pteridines (1) during treatment with dihydrofolate reductase inhibitors was not yet found in urine, neither

treatment with methotrexate (low dose) (7) or cotrimoxazole (7, 10) raised neopterin or biopterin excretion.
The most serious limitations for neopterin determination in neoplastic disease seem so far not the "false negative" results in a great proportion of tumours but its elevation in viral disease, which is frequent in the immunocompromised cancer patient, and in renal insufficiency, which too is frequent in advanced cancers and in elderly people. Therefore viral infections - neopterin levels in trivial illnesses, as common cold, return to normal fairly soon - and renal insufficiency have to be carefully excluded for cancer diagnosis by elevated neopterin levels.
One of the interesting aspects emerging from the studies on pteridines in various diseases concerns the ratio of different pteridines. Syntheses of neopterin and monapterin seem to be connected, since a fairly constant monapterin/neopterin ratio was observed independent of the amount of neopterin excreted. The biopterin content of body fluids is apparently under stricter metabolic control than the neopterin content. Neopterin levels in various diseases vary in a much wider range (Tables 3, 4, 6) than biopterin levels in urine as well as in serum (8). Biopterin is synthesized predominantly in liver, kidney and brain (23-26). Results concerning biopterin biosynthesis were obtained mostly in rats, where no neopterin was found yet. In patients with affected liver and kidney, e.g. renal insufficiency, severe liver failure or liver metastases, neopterin/biopterin ratios in urine were often higher than normal, in some neurological diseases CSF content of biopterin was lowered (12-14). Investigations of neopterin and biopterin levels simultaneously in suitable model systems seem possible now and may yield information on regulation of biosynthesis of neopterin as well as of biopterin.

Acknowledgements

We thank Prof. R. Lamerz, München, for the sera and CEA data, Prof. M. Hüfner, Heidelberg, for his preliminary results, Prof. W. Pfleiderer and Dr. W. Hutzenlaub, Konstanz, Prof. J.H. Bieri, Zürich, and Prof. S. Matsuura, Nagoya, for samples of pteridines.
Thanks are due to Mrs. L. Meister, Ms. B. Zorn for technical assistance and to Mrs. H. Luczak for secreterial help.

References

1. Leeming, R.J., Blair, J.A., Melikian, V., O'Gorman, D.J.: J. Clin. Pathol. 29, 444-451 (1976).
2. Kokolis, N., Ziegler, I.: Cancer Biochem. Biophys. 2, 79-85 (1977).
3. Stea, B., Backlund, P.S., Berkey, P.B., Cho, A.K., Halpern, B.C., Halpern, R.M., Smith, R.A.: Cancer Res. 38, 2378-2384 (1978).
4. Albrecht, A.M., Biedler, J.L., Baker, H., Frank, O., Hutner, S.H.: Res. Commun. Chem. Pathol. Pharmacol. 19, 377-380 (1978).
5. Stea, B.: Dissert. Abstr. Internat. 40, 734 (B) (1979).
6. Wachter, H., Hausen, A., Grassmayr, K.: Hoppe-Seyler's Z. Physiol. Chem. 360, 1957-1960 (1979).
7. Rokos, H., Rokos, K., Frisius, H., Kirstaedter, H.-J.: Clin. Chim. Acta 105, 275-286 (1980).
8. Leeming, R.J., Blair, J.A.: Clin. Chim. Acta 108, 103-111 (1980).
9. Stea, B., Halpern, R.M., Halpern, B.C., Smith, R.A.: Clin. Chim. Acta 113, 231-242 (1981).
10. Rokos, K., Frisius, H., Kirstaedter, H.-J., Fleige, H.-E., John, F., Rokos, H.: in CEA, Carcinoembryonales Antigen, und andere Tumormarker, Eds.: Uhlenbruck, G., Wintzer, G., 408-412, Tumor Diagnostik Verlag Leonberg (1981).
11. Hausen, A., Fuchs, D., Grünewald, K., Huber, H., König, K., Wachter, H.: Clin. Chim. Acta 117, 297-305 (1981).
12. Lovenberg, W., Levine, R.A., Robinson, D.S., Ebert, M., Williams, A.C., Calne, D.B.: Science 204, 624-626 (1979).

13. Williams, A., Ballenger, J., Levine, R., Lovenberg, W., Calne, D.: Neurology 30, 1244-1246 (1980).

14. Nagatsu, T., Yamaguchi, T., Kato, T., Sugimoto, T., Matsuura, S., Akino, M., Nagatsu, I., Iizuka, R., Narabayashi, H.: Clin. Chim. Acta 109, 305-311 (1981).

15. Nagatsu, T., Yamaguchi, T., Kato, T., Sugimoto, T., Matsuura, S., Kobayashi, K., Akino, M., Tsushima, S., Nakazawa, N., Ogawa, H.: Proc. Japan. Acad. 55, Ser. B, 317-322 (1979).

16. Nagatsu, T., Yamagushi, T., Kato, T., Sugimoto, T., Matsuura, S., Akino, M., Tsushima, S., Nakazawa, N., Ogawa, H.: Anal. Biochemistry 110, 182-189 (1981).

17. Fukushima, T., Nixon, J.C.: In Methods of Enzymology Vol. 66, Eds.: McCormick, D.B. and Wright, L.D., 429-436, Academic Press, New York (1980).

18. Fukushima, T., Nixon, J.C.: Anal. Biochemistry 102, 176-188 (1980).

19. Niederwieser, A., Curtius, H.-Ch., Gitzelmann, R., Otten, A., Baerlocher, K., Blehová, B., Berlow, S., Gröbe, H., Rey, F., Schaub, J., Scheibenreiter, S., Schmidt, H., Viscontini, M.: Helv. Paediatr. Acta 35, 335-342 (1980).

20. Dhondt, J.-L., Ardouin, P., Hayte, J.-M., Farriaux, J.-P.: Clin. Chim. Acta 116, 143-152 (1981).

21. Röthler, F., Karobath, M.: Clin. Chim. Acta 69, 457-462 (1976).

22. Hüfner, M.: in preparation.

23. Fukushima, K., Eto, I., Mayumi, T., Richter, W., Goodson, S., Shiota, T.: In Chemistry and Biology of Pteridines, Ed. Pfleiderer, W., 247-263, de Gruyter, Berlin (1975).

24. Häberle, D.A., Schiffl, H., Mayer, G., Hennings, G., Rembold, H.: Pflügers Arch. 375, 9-16 (1978).

25. Kapatos, G., Kaufman, S., Weller, J.L., Klein, D.C.: Science 213, 1129-1131 (1981).

26. Abou-Donia, M.M., Viveros, O.H.: Proc. Natl. Acad. Sci. USA 78, 2703-2706 (1981).

3

PTERIDINES IN CANCER DETECTION AND MONITORING

PTERIDINES IN URINE AND SERUM FROM CANCER PATIENTS

J.L. DHONDT, J.M. HAYTE

Laboratoire de biochimie, Faculté libre de Médecine, 56 rue du port, 59046 Lille Cédex (France)

J. BONNETERRE, L. ADENIS and A. DEMAILLE

Centre Oscar Lambret, 59000 Lille (France)

P. ARDOUIN, J.P. FARRIAUX

Service de Génétique et Maladies Héréditaires du Métabolisme. Cité Hospitalière, 59037 Lille Cédex (France)

Several recent reports have demonstrated the occurrence of elevated levels of certain unconjugated pteridines in urine of cancer patients (1-3). However, these data concerned various types of malignant diseases, thus no significant conclusion could be offered concerning the frequency and the mechanism of these abnormalities.

The aim of this work was to investigate pteridine metabolism in two homogenous groups of cancer patients : a male group with cancers of the upper respiratory tract (neck cancers) and a female group with breast cancers, by studying pteridine levels in serum and urine.

Material and Method :

Controls : Urines were collected from 54 healthy controls (38 males and 15 females) and blood from 26 controls (10 males and 16 females).

Patients : 47 patients with neck cancers (aged 39-70 years) and 64 patients with breast cancer (aged 32-80 years) were studied. None of them received cytotoxic or hormonal therapy at the time of analysis.

Method : Urines and/or sera were obtained mostly in the

Biochemical and Clinical Aspects of Pteridines · Volume 1
Cancer · Immunology · Metabolic Diseases

morning. Urines were immediately acidified to pH 1 with 1 N HCl, protected from light and stored at -30°C. Sera were immediately separated following the vein puncture and stored at -30°C.

Isolation and chromatography of pteridines were performed according to the method previously described (4). The pteridines (1 ml of urine, 2 to 5 ml of serum) were oxidized in their more fluorescent derivatives by iodine in 0.1 N HCl for 1 hour. Excess iodine was then reduced by ascorbic acid and the internal standard (6-methylpteridine) was added : 10 µg in urine samples and 0.4 µg in serum. Sera were then deproteinized by trichloroacetic acid (final concentration 0.4 M) and centrifuged. The samples were purified through ion-exchange chromatography with Dowex 50W H^+ and Dowex 1 OH^-. The final acetic eluates : 2 ml from urine, 0.5 ml from serum, were directly analyzed by high performance liquid chromatography, using a cation exchange column (Partisil 10-SCX, Whatman) and a fluorimetric detector as described in (4).

In regard to the low serum concentration of pteridines, xanthopterin and pterin were not quantified in sera.

Results :

The mean urinary excretion and serum level of pteridines of healthy adults and cancer patients are shown in table I and in figure 1. In controls, no significant differences between the sexes have been noted in the pteridine levels in serum and urine.

In neck cancers, a significant increase of the elimination of xanthopterin, neopterin and pterin, and of the value of the neopterin to biopterin ratio (N/B) was observed when compared to controls. When

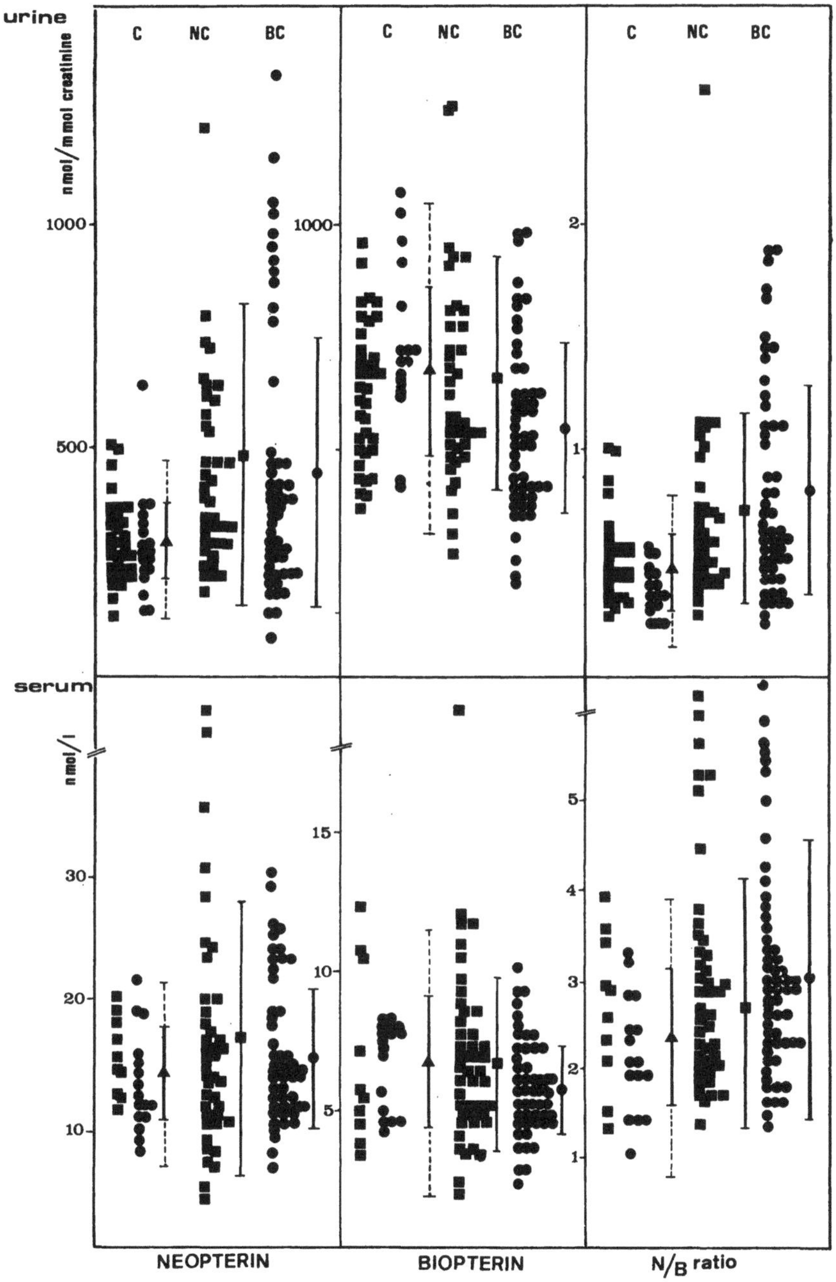

Figure 1

distibution of pteridines in urine and serum of healthy and cancer patients
■ males, ● females, ▲ both sexes(mean ± 1 SD, dashed line: mean ± 2 SD)
C: controls, NC: neck cancers, BC: breast cancers.

Table I:

Concentration of pteridines in urine and serum from controls and cancer patients.

	Xanthopterin	Neopterin	Biopterin	Pterin	N/B ratio
URINES: (nmol/mmol of creatinine)					
Controls: (n=53)	193 ± 92	309 ± 91	679 ± 186	64 ± 25	0.48 ± 0.18
Neck cancers: (n=40)	367 ±152 $p<0.001$	500 ±344 $p<0.001$	672 ± 260 ns	99 ± 51 $p<0.001$	0.78 ± 0.45 $p<0.001$
Breast cancers: (n=54)	340 ±224 $p<0.001$	474 ±312 $p<0.001$	554 ± 191 $p<0.001$	110 ± 68 $p<0.001$	0.87 ± 0.48 $p<0.001$
SERUM: (nmol/l)					
Controls: (n=26)	nd	15.4 ± 4.1	6.8 ± 2.4	nd	2.4 ± 0.8
Neck cancers: (n=47)	nd	17.9 ±11.8 ns	6.7 ± 3.2 ns	nd	2.7 ± 1.4 ns
Breast cancers: (n=57)	nd	16.0 ± 1.6 ns	5.5 ± 1.6 $p<0.01$	nd	3.0 ± 1.6 ns

nd : not determined
ns : not significative

individual values were considered, the N/B ratio was increased (greater than mean normal value + 2 SD) in 25% of the cases. In contrast, no significant change in serum was noted; however, analysis of individual values revealed high neopterin levels in 17% and an elevated N/B ratio in 15% of the cases.

In breast cancers, the mean xanthopterin, neopterin and pterin excretion was significantly increased, whereas biopterin elimination was decreased. The mean N/B ratio was therefore higher than in the previous group and elevated individual values were noted in 39% of the cases. In serum, only a significant decrease of biopterin concentration was observed, but 19% of the cases had high neopterin levels and elevated N/B ratio.

In the both groups of cancer, any correlation between pteridine levels and either cellular typing or metastatic potential of the tumors could be found.

Discussion :

The pathway of the biosynthesis of tetrahydrobiopterin has been in part elucidated (5-8). Issued from Guanosine-triphosphate, dihydroneopterin-triphosphate is transformed into dihydrobiopterin by a Biopterin-synthetase complexe. Then dihydrobiopterin is reduced by dihydrofolate-reductase into tetrahydrobiopterin, which serves as a cofactor of the hydroxylation of phenylalanine, tyrosine and tryptophan.

Altered metabolism of pteridines has been demonstrated in animals with experimental tumors (9) and in cancer cells in culture (10-12). Abnormal elimination of pteridines in urine of cancer patients was reported by Rokos (1), Wachter (2) and Stea (3). However, the wide variety of studied cancers, the different therapeutic conditions and the small size of patient groups limited the interpretation of these results. Moreover, no information was available about the concentration of pteridines in the serum of these patients.

Testing two of the most frequent cancerous diseases, we confirmed the increased excretion of neopterin in cancer, but with a frequency lower than that previously speculated (1-3). In contrast to results found in urine, pteridine concentrations in serum were not significantly different from those of controls, excepted the slight decrease of biopterin levels in breast cancer. However, when individual values were considered, elevated concentration of neopterin and/or high N/B ratio were observed in 15-19% of the cases. When plotting serum N/B ratio versus urinary N/B ratio, the regression curve was statistically significant ($n=96$, $r=0.68$, $p < 0.001$).

The main abnormality in the distribution of pteridines in biological fluids of cancer patients was thus an increased N/B ratio. This finding was of some interest in regard to the variations of this ratio during the maturation (13,14). In fact the N/B ratio progressively decreased, in urine: from 4.4 ± 2.1 in newborns (n=55) to 0.5 ± 0.2 in adults, and in serum: respectively from 11.4 ± 7.5 (n=13) to 2.4 ± 0.8 (14), suggesting a progressive maturation of the pteridine metabolism. Thus the high N/B ratio found in some cancer could be related to the dedifferentiation process occuring in cancer tissues.

The low frequency of such abnormalities could be explained either by different metabolic activities of the tested cancers or by differences in their size. Moreover, the diagnostic value of these abnormalities remained poor regarding the possibilities of similar changes in other pathological situations such as kidney dysfunction (1) or virus diseases (2). Though, from a basic point of view, these disturbances in cancer are of some interest. In fact, tetrahydrobiopterin and some of the enzymes involved in its metabolism are known to be present in many tissues, some of them being devoid of any of the hydroxylases which require tetrahydrobiopterin as a cofactor. Thus recent speculations have arisen about other metabolic roles of tetrahydrobiopterin (15-17). The description of raised concentration of pteridines in biological fluids of cancerous subjects and of altered enzymatic activities of this metabolism in some tumors (18) could lead to the use of cancer tissues as a model for the approach of new metabolic roles of tetrahydrobiopterin and of regulation of pteridine metabolism.

References:

1. Rokos,H., Rokos,K., Frisius,H. and Kirstaedter, H-J. Alterated urinary excretion of pteridines in neoplastic disease. Determination of biopterin, neopterin, xanthopterin and pterin. Clin. Chim. Acta 105, 275-286 (1980).

2. Wachter,H., Hausen,A. and Grassmayr,K. Erhöhte Ausscheidung von Neopterin in Harn von Patienten mit malignen Tumoren und mit Viruserkrankungen. Hoppe Seyler's Z. Physiol. Chem. 360, 1957-1960 (1979).

3. Stea,B., Halpern,R.M., Halpern,B.C. and Smith,R.A. Urinary excretion levels of unconjugated pterins in cancer patients and normal individuals. Clin. Chim. Acta 113, 231-242 (1981).

4. Dhondt,J.L., Largilliere,C., Ardouin,P., Farriaux,J.P. and Dautrevaux,M. Diagnosis of variants of hyperphenylalaninemia by determination of pterins in urine. Clin. Chim. Acta 110, 58-60 (1981).

5. Fukushima,K., Richter,W.E. and Shiota,T. Partial purification of 6-(D-erythro-1',2',3'-trihydroxypropyl)-7,8-dihydropterin triphosphate synthetase from chicken liver. J. Biol. Chem. 252, 5750-5755 (1977).

6. Gal,E.M., Nelson,J.M. and Sherman,A.D. Purification and characterization of enzymes involved in the cerebral synthesis of 7,8-dihydrobiopterin. Neurochem. Res. 3, 69-88 (1978).

7. Tanaka,K., Akino,M., Hagi,Y., Doi,M. and Shiota,T. The enzymatic synthesis of sepiapterin by chicken kidney preparations. J. Biol. Chem. 265, 2963-2972 (1981).

8. Häusermann, M., Ghisla, S., Niederwieser, A. and Curtius, H.Ch. New aspects in biopterin biosynthesis in man. FEBS Letters 131, 275-278 (1981).

9. Kokolis, N. and Ziegler, I. On the levels of phenylalanine, tyrosine and tetrahydrobiopterin in the blood of tumor-bearing organisms. Cancer Biochem. Biophys. 2, 79-85 (1977).

10. Albrecht,A.M., Biedler,J.L., Baker,H., Frank,O. and Hutner,S.H. Biopterin content of human neuroblastoma cells in culture. Res. Commun. Chem. Path. Pharm. 19,377-380 (1978).

11. Stea,B., Backlund,P.S., Berkey,P.B., Cho,A.K., Halpern,B.C., Halpern,R.M. and Smith,R.A. Folate and pterin metabolism by cancer cells in culture. Cancer Res. 38, 2378-2384 (1978).

12. Katoh, S., Sueoka, T. and Yamada, S. Studies on the cofactor of phenylalanine hydroxylase in some transplantable rat ascites hepatomas. in Kisliuk, R.L. and Brown, G.M. Chemistry and Biology of Pteridines (Elsevier, New York 1979) p.189-194.

13. Niederwieser,A., Curtius,H-Ch., Gitzelmann,R., Otten,A., Baerlocher,K., Blehova,B., Berlow,S., Gröbe,H., Rey,F., Schaub,J., Scheibenreiter,S., Schmidt,H. and Viscontini,M. Excretion of pterins in phenylketonuria and phenylketonuria variants. Helv. Paed. Acta 35, 335-342 (1980).

14. Dhondt,J.L., Ardouin,P., Hayte,J.M. and Farriaux,J.P. Developmental aspects of pteridine metabolism and relationships with phenylalanine metabolism. Clin. Chim. Acta 116, 143-152 (1981).

15. Rembold,H. and Buff,K. Tetrahydrobiopterin, a cofactor in mitochondrial electron transfer. Effect of tetrahydropterins on intact rat-liver mitochondria. Eur. J. Biochem. 28, 579-585 (1972).

16. Taylor, D. and Hochstein, P. Reduction of methemoglobin by tetrahydropterin and glutathione. Arch.Biochem.Biophys. 179, 456-461 (1977).

17. Snyder,F., Malone,B. and Piantadosi,C. Tetrahydropteridine-dependent cleavage enzyme for O-alkyl lipids: substrate specificity. Biochem. Biophys. Acta 316, 259-265 (1973).

18. Dhondt,J.L., Bonneterre,J., Farriaux,J.P., Lefebvre,J. and Demaille,A. Dihydropteridine-reductase activity in breast cancer. Preliminary report. Biomedicine 35, 58-60 (1981).

SIGNIFICANCE OF URINARY NEOPTERIN IN GYNECOLOGY AND OBSTETRICS: ONCOLOGY, PREGNANCY, AMNIOTIC FLUID

Alfred Bichler, Heinz Hetzel, Christian Marth
Universitätsklinik für Frauenheilkunde Innsbruck
A-6020 Innsbruck, Austria

Dietmar Fuchs, Arno Hausen, Gilbert Reibnegger and Helmut Wachter
Institut für Medizinische Chemie und Biochemie der Universität Innsbruck
A-6020 Innsbruck, Austria

Introduction

In recent years the possible role of pteridines in the metabolism of cancer cells has been investigated. Enhanced pteridine levels were found in cultures of cancer cells (1), tumor-bearing animals (2,3) and in humans suffering from cancer (4-9). The studies in humans are very heterogeneous as they comprise a broad range of various types and stages of cancer in women and men. Hence our investigations on urinary neopterin (6-trihydroxypropylpterine) excretion in cancer were confined to women with carcinoma of the genitals. Figure 1 shows a scheme of frequency and typical localisation of cancer of the female genital tract. These tumors differ by morphological and histological criteria. The precancerous stages of some of these cancers

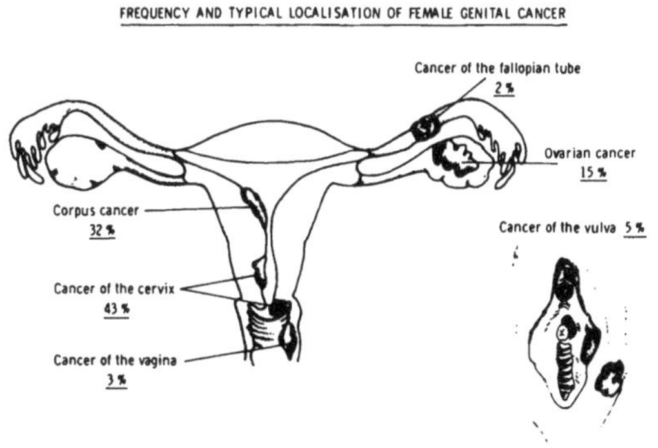

Fig. 1

Biochemical and Clinical Aspects of Pteridines · Volume 1
Cancer · Immunology · Metabolic Diseases

are exactly defined and easily diagnosed. Hence we were able to correlate neopterin excretion with different types and stages of genital cancer.
The theoretical basis for the elevated excretion of some pteridines in cancer patients is poorly understood. The purpose of this paper was to help elucidate the role of neopterin in the human by monitoring it's excretion in patients with various benign and malignant gynecological tumors and in altered normal conditions such as pregnancy.

Patients and methods

Urinary neopterin determinations were performed:

1. on a group of apparently healthy women (n=209) and men (n=208)
2. on 53 patients with various beningn gynecological tumors
3. on 24 women with precancerous lesions of the cervix and the endometrium
4. as serial determinations over a period of one and a half year on 108 patients with cancer of the genitals
5. on 218 pregnant women at varying times of gestation and on 20 puerperae
6. in 39 amniotic fluid samples at varying times of gestation

In each sample neopterin was estimated simultaneously with creatinine. The values of neopterine were related to the creatinine concentration in the urine.
The HPLC-method of neopterin determination in human urine is described by Fuchs in these proceedings (10). The quantitative estimation of creatinine was also performed by HPLC (11).
In almost all cases urine collected in the morning was used for the determinations. Specimens contaminated with blood were discarded. The samples were analysed either immediately or stored at -20°C in the dark. All analyses were performed without knowledge of the clinical diagnosis.

Statistical analysis (12):

The Wilcoxon and Mann-Whitney tests as well as the Student-t-test (checking for normal distribution) were used for comparing the mean values. Correlations were calculated with the Pearson coefficient of correlation. Heterogeneous distributions were checked by the X^2 test. The limits of tolerance were tested according to Liebermann.

Results

Neopterin values in healthy women and men

neopterin (µ mol/mol creatinine)

	healthy men n = 208		healthy women n = 209		
Age (years)	n	$\bar{x} \pm SD$	n	$\bar{x} \pm SD$	p
18 - 25	42	123 ± 30	55	128 ± 33	n.s.
26 - 35	29	101 ± 33	28	124 ± 33	< 0.01
36 - 45	41	109 ± 28	31	140 ± 39	< 0.001
46 - 55	32	105 ± 36	28	147 ± 32	< 0.001
56 - 65	31	119 ± 39	26	156 ± 35	< 0.005
> 66	33	133 ± 38	41	151 ± 40	n.s.

Mean urinary neopterin values (µ mol/mol creatinine) and standard deviation in healthy women and men.

Tab. 1

Table 1 shows the median urinary neopterin values (µmol/mol creatinine) in healthy women and men in different age groups. In women, significantly higher values were found than in men in almost all age groups. Furthermore there was a weak positive correlation between the neopterin /creatinine ratio and age, both in men (r=0,159, $p < 0,05$), and women (r=0,217, $p < 0,01$). This seems to be due mainly to a sex and age dependent alteration of the urinary creatinine concentration. Significantly lower creatinine values were found in women than in men ($p < 0,002$) as well as a negative correlation between urinary creatinine values and age both in men (r=-0,376, $p < 0,0001$) and women (r=-0,380, $p < 0,0001$).

The upper confidence limits of the neopterin values were established for each age group of healthy women according to Liebermann (12). With a 95% probability, 95% of our healthy women

Age (years)	n	neopterin (µmol / mol creatinine)
18 - 25	55	202
26 - 35	28	202
36 - 45	31	230
46 - 55	28	222
56 - 65	26	235
> 66	41	243

Upper confidence limits of the urinary neopterin values in healthy women (with a 95% probability, 95% of our healthy women are within these limits).

Tab. 2

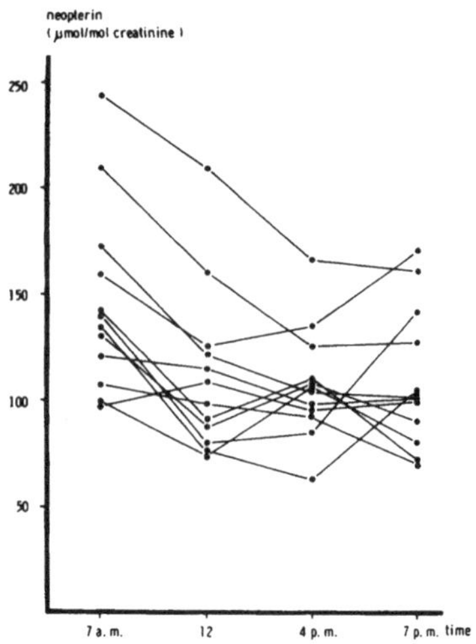

Daily variation of the urinary neopterin excretion in young healthy women (n = 12)

Fig. 2

are within these limits (table 2). In figure 2 the daily variation of the urinary neopterin excretion in 12 young (mean age=21 years) healthy women is shown. There is a trend towards decreasing levels in the evening but it is not significant at the 5% level.

Figure 3 shows the monthly variation of the neopterin excretion in morning urine samples of a healthy women with biphasic cycle. No difference can be seen in neopterine excretion between the first and second half of the menstrual cycle. The coefficient of variation was 41%.

<u>Neopterin values in patients with benign gynecological tumors</u>

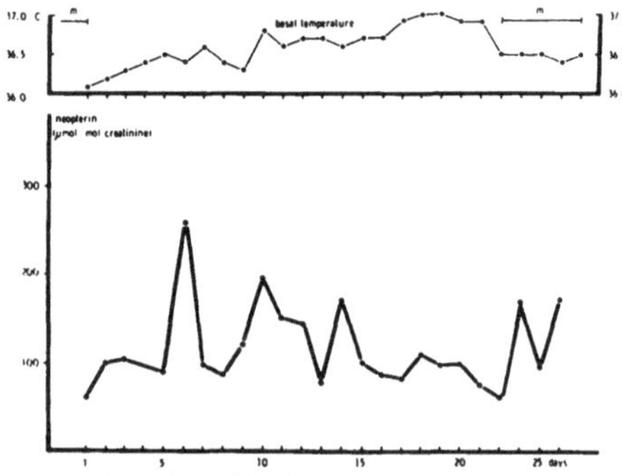

Lower figure: Monthly variation of the neopterin excretion in morning urine samples of a healthy women (K. M., 23 years)
Upper figure: Basal temperature curve of this woman.

Fig. 3

Urinary neopterin determinations were done in 53 patients with benign gynecological tumors (myomas, ovarian cysts).

When compared to the neopterin values of healthy women of the same age groups (36-45 and 46-55 years), no significant difference was seen (table 3). In the other age groups the number of patients with benign gynecological tumors was too small for a statistical comparison.

neopterin (μmol/mol creatinine)

Age (years)	healthy women n	$\bar{x} \pm SD$	benign gynecological tumors n	$\bar{x} \pm SD$	p
36 - 45	31	140 ± 39	22	167 ± 59	n.s.
46 - 55	28	147 ± 32	15	192 ± 85	n.s.

Mean urinary neopterin values (μmol / mol creatinine) and standard deviation in healthy women and in patients with benign gynecological tumors.

Tab. 3

Neopterin values in patients with precancerous lesions of cervix and endometrium

Urinary neopterin was sampled in 24 patients with precancerous lesions of the cervix (dysplasia, carcinoma in situ) and of the endometrium (adenomatous hyperplasia). The small number of patients did not allow a statistical comparison with the healthy control group. However, 19 out of these 24 patients had neopterin levels below the upper confidence limits of normal values. Thus we feel that there is no difference in neopterine levels between healthy women and patients with precancerous lesions of the cervix and the endometrium.

Neopterin values in patients with genital cancer

Significantly higher mean urinary neopterin levels were found in 108 patients with cancer of the genitals before treatment, when compared to the healthy control group (table 4).

The significance of neopterin sampling in monitoring each single patient with cancer is shown in the tables 5,6 and 7. The upper confidence limits of the normal neopterin values (see table 2) were used for the evaluation.

In table 5 the significance of neopterin values in patients

neopterin (μ mol/mol creatinine)

	healthy women n = 154		cancer of the genitals n = 108		
Age (years)	n	$\bar{x} \pm SD$	n	$\bar{x} \pm SD$	p
26 - 35	28	124 ± 33	13	203 ± 99	< 0.02
36 - 45	31	140 ± 39	18	200 ± 100	< 0.05
46 - 55	28	147 ± 32	7	303 ± 167	< 0.005
56 - 65	26	156 ± 35	34	320 ± 183	< 0.001
> 66	41	151 ± 40	36	361 ± 178	< 0.001

Mean urinary neopterin values (μ mol/mol creatinine) and standard deviation in healthy women and in patients with cancer of the genitals prior to the beginning of therapy.

Tab. 4

	n	correspondence
ovarian cancer	15	14
cervical cancer	66	31
endometrial cancer	24	14
cancer of the vulva	3	2
total	108	61 56%

Significance of urinary neopterin values (μ mol/mol creatinine) prior to the beginning of therapy in patients with cancer of the genitals.

Tab. 5

with different types of genital cancer prior to the beginning of therapy is demonstrated. In 56% of all patients with cancer elevated neopterin levels were found. As seen in table 5 the best correspondence was found in patients with ovarian cancer (93%) whereas for patients with cervical cancer agreement was only 47%.

These 108 patients underwent treatment (operation, irradiation, cytostatic therapy) and some of them were followed up one and a half year by neopterin sampling. The clinicoroentgenological findings in these patients were compared with the urinary neopterin values. Results are shown in the tables 6 and 7. In 81% of cases of pretherapeutic elevated neopterin levels the clinicoroentgenological findings for detection of cancer (palpation, sonography, lymphography, computertomography) corresponded with neopterin values. In 19% false positive neopterin levels were found, i.e. there were elevated neopterin values but no clinicoroentgenological signs of recurrence or progression of the carcinoma. False negative neopterin levels (normal values but recurrence or progression of the malignant tumor) were never seen. The highest rate of correspondence seems to be in patients with ovarian cancer (table 6). However, it was possible in only about 50% of the patients with

	before therapy elevated neopterin values n	correspondence n	follow-up false pos. n	false neg. n
ovarian cancer	9	9	-	-
cervical cancer	16	13	3	-
endometrial cancer	6	3	3	-
total	31	25 81%	6 19%	- -

Urinary neopterin concentration (μmol/mol creatinine) and clinicoroentgenological findings in the follow-up of the patients with pretherapeutic elevated neopterin values.

Tab. 6

	before therapy normal neopterin values n	follow-up correspondence n	false pos. n	false neg. n
ovarian cancer	1	1	-	-
cervical cancer	27	21	5	1
endometrial cancer	3	3	-	-
cancer of the vulva	1	1	-	-
total	32	26 81%	5 16%	1 3%

Urinary neopterin concentration (μmol/mol creatinine) and clinicoroentgenological findings in the follow-up of the patients with pretherapeutic normal neopterin values.

Tab. 7

pretherapeutic elevated neopterin levels to follow by neopterin sampling over a period of one and a half year.

The significance of neopterin sampling in the follow-up of the patients with pretherapeutic normal values is shown in table 7.

Our data don't allow to conclude if elevated urinary neopterin excretion is an early sign of tumor progression or recurrence in the follow-up of cancer patients. This has to be studied in detail in further series.

An attempt was made to correlate the neopterin values with stage and grade of the malignant tumors.

We couldn't find any stage dependent change in the neopterin excretion in our patients with cancer of the genitals. Thus it seems that the size of these malignant tumors doesn't influence the neopterin excretion.

In almost all cervical cancers highly differentiated cells were found whereas in most of the ovarian carcinomas undifferentiated cells were seen. In endometrial cancer various grades of differentiation were observed. The greatest frequency of elevated neopterin levels was found in patients with ovarian cancer (14 from 15), whereas the lowest rate was observed in women with cervical cancer.(31 from 66). Up to now no correlation between the differentiation of the tumor (grade 1-3) and the neopterin values has been found in patients with endometrial adenocarcinoma.

Neopterin values during gestation and in the puerperium

As shown in figure 4 increasing mean urinary neopterin levels were found with advancing normal gestation but the difference between these three periods is not significant. During the

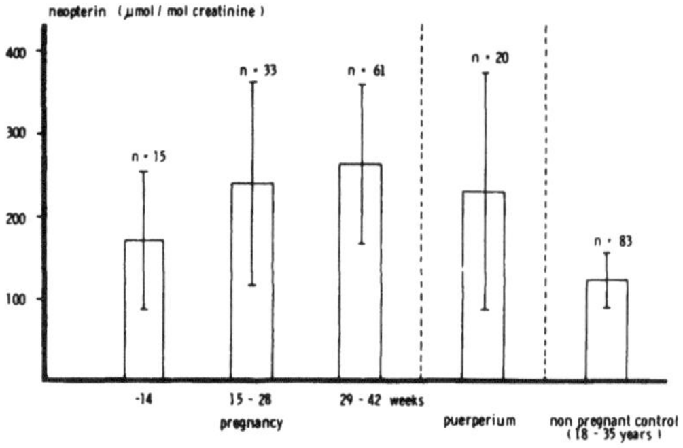

Mean urinary neopterin values (μmol / mol creatinine) and standard deviation during normal pregnancy and in the puerperium.

Fig. 4

early puerperium neopterine values found were similar to those of the second period of pregnancy. However, in pregnancy and in the early puerperium the neopterin excretion was significantly higher than in a non pregnant control (figure **4**).

In eight women carrying twins neopterin values were twofold higher than in patients with single pregnancies.

Urinary neopterin was also measured in 109 pathological pregnancies as EPH-gestosis, threatened abortion or premature delivery, diabetes mellitus, postdated pregnancy. As compared to normal pregnancies, in some pathological pregnancies elevated neopterin levels were found. However, we cannot say from this first serie of high risk pregnancies on which typical pathological patterns an altered neopterin excretion occurs.

Neopterin values in the amniotic fluid

On a small number of patients (n=39) neopterin was estimated in the amniotic fluid. Indications for amniotic fluid sampling by amniocentesis were mostly Rh-sensitization.

When plotted against number of weeks of gestation, our first results show an increase in amniotic fluid neopterine concentration after the 34th week (figure 5). There is a strong correlation between amniotic fluid neopterin and creatinine concentration ($r=0.7707$, $p < 0.0001$) throughout the course of pregnancy (figure 5).

Case report. Miscellaneous

Figure 6 shows the course of a 38 years old patient with chorionic carcinoma of the uterus. Neopterin excretion was compared to the specific markers for this type of malignant tumor. As seen in figure 6 ,urinary neopterin was elevated

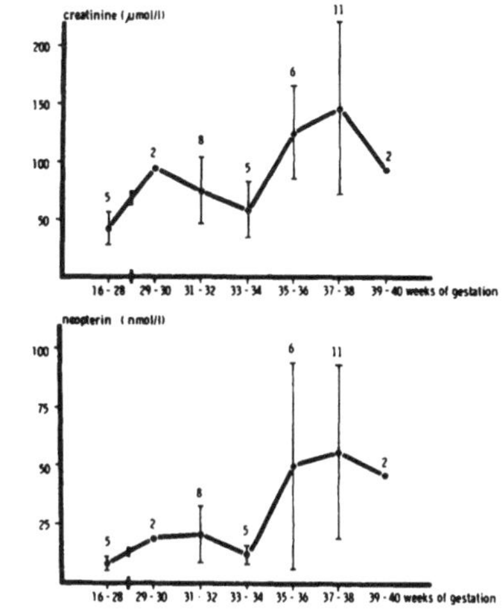

Neopterin and creatinine concentrations in the amniotic fluid ($\bar{x} \pm SD$)

Fig. 5

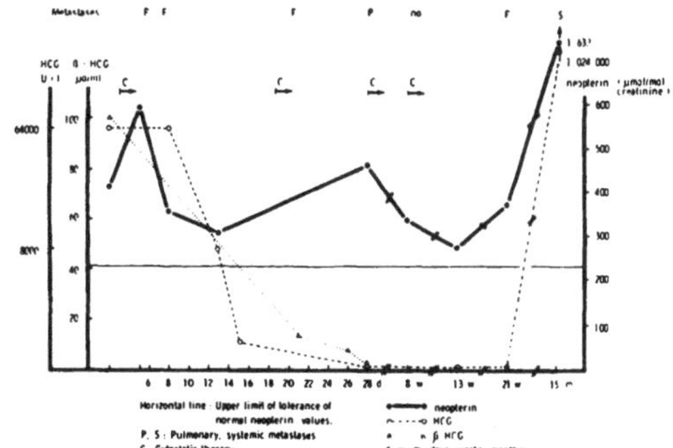

N. M., 38 years. Chorionic carcinoma of the uterus with formation of metastases. Surgical treatment and serial cytostatic therapy preceded this period of observation.

Fig. 6

throughout the period of observation whereas the HCG- and ß-HCG-concentrations dropped to normal values with tumor regression and disappareance of the metastases. Tumor progression, which occured within a few weeks, was accompanied by a sharp increase in HCG as well as neopterine excretion. The patient died some days later because of the severely progressed carcinoma.

Like Wachter et al. (7,13), we found elevated neopterin levels in patients with viral infections and during the first five days after surgical procedure.

These observations are illustrated exemplary in the figures 7 and 8.

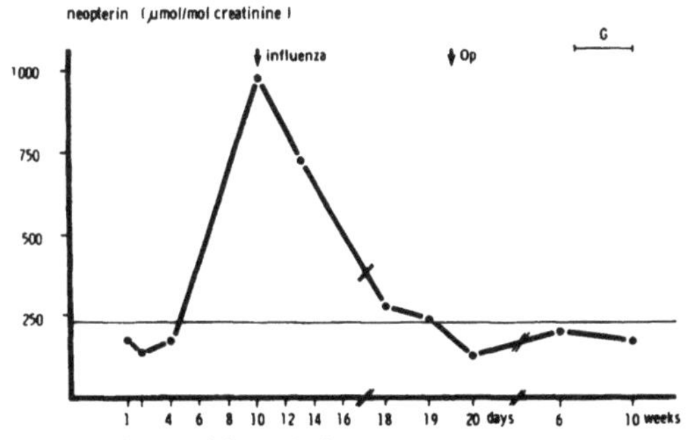

M. A., 62 years, cervical cancer stage Ib
Op.: Latzko-operation
Horizontal line: Upper limit of tolerance of normal neopterin values. G: Gamma irradiation

Fig. 7

Discussion

Neopterin (6-trihydroxypropylpterine) belongs to the pteridines which are ubiquitous (14).

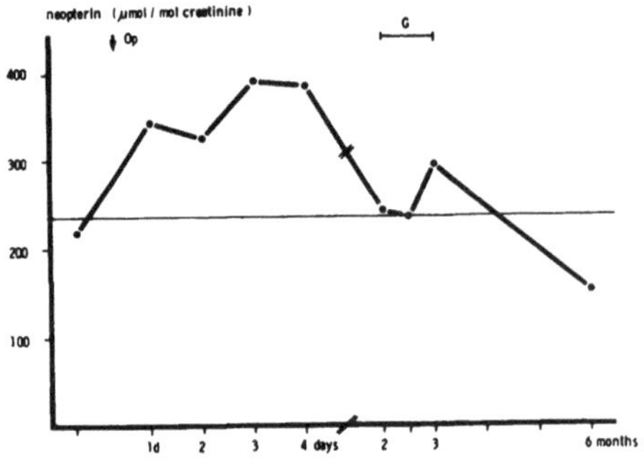

Fig. 8

The pteridines play a role in photosynthesis and in the cellular electron transfer system (15,16). In their reduced form they are required as cofactors for the hydroxylation of phenylalanine, tyrosine and tryptophane (17,18).

In recent years the possible role of pteridines in cancer metabolism in animals, cell cultures and in humans has been investigated. Enhanced levels of tetrahydrobiopterin were shown in the blood of cancer patients and of tumor bearing rats (2). Mice with Ehrlich ascites tumors were found to excrete increased quantities of 7,8-dihydro-6-hydroxylumazine (3), a pteridine, which is probably converted from neopterin by deaminases.

Cancer cells were shown to excrete in their growth medium 6-hydroxymethylpterin (1). The authors suggested that the production of this pteridine could indicate active folic acid catabolism by malignant cells.

Clinical reports have given some evidence of altered pteridine excretion in cancer patients. In our laboratory elevated neopterin levels have been found in patients with various malignant tumors (4,7,8,9). Similar results were shown by Rokos (5) and by Stea (6).

Our results confirm these clinical reports. Significantly higher mean urinary neopterin values were found in patients with cancer of the genitals as compared to a healthy control group. We were unable to show that neopterin is elevated in women with benign gynecological tumors and in patients with precancerous lesions of the cervix and the endometrium.

The reasons for the increased excretion of some pteridines

in patients with malignant tumors are still poorly understood. It was suggested that pteridine excretion is paralleled with cell proliferation as elevated levels were found in E.coli-cultures during cell proliferation (13) and during tail regeneration in Triturus species (19). The elevated urinary neopterin levels during the first days after a surgical procedure as well as the increasing levels with advancing gestation give additional support for a relation between cell proliferation and pteridine excretion. On the other hand, enhanced cell proliferation cannot be the only reason for increased pteridine excretion, as normal neopterin values were found in patients with benign gynecological tumors. In addition, the size of the malignant tumor doesn't correlate with the neopterine excretion.

The pteridine levels in cancer patients could be influenced by the grade of differentation of the tumor. On ovarian carcinomas, which are generally undifferentiated tumors, the greatest frequency of elevated neopterin levels was found whereas the lowest rate was observed in women with highly differentiated squamous cell carcinomas of the cervix. These data agree with the results of neopterin sampling in patients with squamous cell carcinomas of the larynx (20). In this highly differentiated type of cancer, normal values were found in almost all cases.

Very recent data from this laboratory suggest that activation of T lymphocytes with alloantigens or with viral antigens is associated with increased production of pteridines (21). These findings suggest new aspects of the significance of pteridines e.g. in graft immunology (22), in cancer metabolism and in pregnancy.

In pregnancy neopterin sampling could be useful in these cases where immunological disturbances are discussed to be responsables for the pathological conditions, i.e. in EPH-gestosis, habitual abortion and in the hydatide mole. This has to be studied more extensively as in our first serie of neo -

pterin excretion in pathological pregnancies this special group of patients wasn't observed in a representative number. The estimation of the neopterine concentration in the amniotic fluid may open some new interesting aspects at the assessment of the fetus. Our first results suggest that the amniotic fluid neopterin concentration is increased after the 34th week of gestation. It is striking that the well established tests for estimating fetal maturity as amniotic fluid creatinine and lecithin concentrations show a similar curve during gestation as neopterin . Neopterin in the amniotic fluid may derive from the mother by diaplacental passage or (and) from the fetus by his urine. This has to investigated in order to establish the value of amniotic fluid neopterin sampling as a method of the fetal assessment as maturity, malnutrition or immunological intolerance.

In conclusion, our results and those of others (4-9) convince us that neopterin sampling is an additional help in monitoring patients with some types of cancer. In gynecology, we have shown that assessment of neopterin seems to be useful in the detection and especially in the follow-up of patients with ovarian cancer. Furthermore, in this type of cancer there is some need for additional methods of early detection of recurrence or progression of the tumor. In patients with cancer of the cervix or of the endometrium neopterin sampling could be suitable only in the follow-up.

The significance of pteridines during pregnancy or in the assessment of the fetus remains to be elucidated.

Summary

Urinary neopterin was sampled in healthy women (n=209) and men (n=208), in patients with benign gynecological tumors (n=53), in women with precancerous lesions of the cervix and the endometrium (n=24) and in patients with cancer of the ge-

nitals (n=108). In addition neopterin determinations were done on 218 pregnant women at varying times of gestation, on 20 puerperae and in 39 amniotic fluid samples at varying times of gestation.
No significant difference was found between the mean neopterin values of patients with benign gynecological tumors or precancerous lesions and the mean values of healthy women. In patients with cancer, significantly higher mean neopterin levels were found than in the control group. Elevated neopterin levels were seen in 56% of the patients with cancer of the genitals varying between 93% (ovarian cancer) and 47% (cervical cancer). Some of the cancer patients were followed up over a period of one and a half year by serial neopterin determinations and clinicoroentgenological examinations. About 80% correspondence between the neopterin values and the the clinicoroentgenological findings of these patients was found, the best agreement being seen on the follow-up of patients with ovarian cancer.
Significantly higher median neopterin values were found during normal pregnancy and in the early puerperium as compared to the non pregnant healthy control.
In the amniotic fluid samples about tenfold lower neopterin values were found as compared to the urine concentrations on the healthy non pregnant women. An increase in the mean amniotic fluid neopterin concentration after the 34th week of gestation was seen.
The reasons for elevated neopterin excretion in patients with cancer and in the pregnancy are poorly understood. Among others, enhanced cell proliferation and alloantigenic activation of T lymphozytes are assumed.
In conclusion, urinary neopterin sampling seems to be an additional help in monitoring patients with cancer of the genitals, especially in the follow-up of patients with ovarian cancer. The significance of neopterin during pregnancy or in the assessment of the fetus by amniotic fluid sampling remains to be elucitated.

References

1. Stea, B., Backlund,Jr. P.S., Berkey, Ph.B., Cho,A.K., Halpern, B.C., Halpern,R.M.,Smith,R.A.: Cancer Research 38, 2378-2384 (1978)

2. Kokolis, N., Ziegler, I.: Cancer Biochem. Biophys. 2, 79-85 (1977)

3. Wachter, H., Grassmayr, K., Hausen, A.: Cancer Lett. 6, 61-66 (1979)

4. Bichler, A., Fuchs, D., Hausen, A., Hetzel, H., König, K., Wachter, H.: Clin. Biochem. 15 (1), 38-40 (1982)

5. Rokos, H., Rokos, K., Frisius, H., Kirstaedter, H.J.: Clin. Chim. Acta 105, 275-286 (1980)

6. Stea, B., Halpern, R.M., Halpern, B.C., Smith, R.A.: Clin. Chim. Acta 113, 231-242 (1981)

7. Wachter, H., Hausen, A., Grassmayr, K.: Hoppe-Seyler's Z. Physiol. Chem. 360, 1957-1960 (1979)

8. Hausen, A., Fuchs, D., Grünewald, K., Huber, H., König, K., Wachter, H.: Clin. Chim. Acta 117, 297-305 (1981)

9. Hausen, A., Fuchs, D., Grünewald, K., Huber, H., Wachter, H.: Clin. Biochem. 15 (1), 34-37 (1982)

10. Fuchs, D., Hausen, A., Reibnegger, G., Wachter, H.: In Biochemical and Clinical Aspects of Pteridines. First

Workshop on Pteridines. St.Christoph, Arlberg (Austria), March 7-1o, 1982, pp 67-79, Eds.: H.Wachter, H.Ch.Curtius, W.Pfleiderer. Walter de Gruyter, Berlin - New York (1982)

11. Hausen, A., Fuchs, D., Wachter, H.: J. Clin. Chem. Clin. Biochem. 19, 375-378 (1981)

12. Sachs, L.: Angewandte Statistik. 5. Auflage, Springer Verlag. Berlin-Heidelberg-New York 1978

13. Wachter, H., Hausen, A., Reider, E., Schweiger, M.: Naturwissenschaften 67, 610-611 (1980)

14. Pfleiderer, W.: Chemistry and Biology of Pteridines. Proc. 5th Int. Symp. Konstanz 1975, Walter de Gruyter, Berlin-New York 1975

15. Fuller, R.S., Nugent, N.A.: Proc. Nat. Acad. Sci USA 63, 1311-1318 (1969)

16. Rembold, H.: In: Iwai, K., Goto, M., Akino, M., Iwanami,Y.: Chemistry and Biology of Pterdines. Proc. 4th Int. Symp. Pteridines Toba 1969, Internat Acad. Printing Co, Tokyo, p 163, 1970

17. Kaufmann, S.: Ann. Rev. Biochem. 36, 171-184 (1967)

18. Lloyd, T., Weiner, N.: Molec. Pharmacol. 7, 569-580 (1971)

19. Kokolis, N., Mylonas, N., Ziegler, I.: Z. Naturforsch. 27b, 285-291 (1972)

20. Reibnegger, G., Bichler, E., Böheim, K., Fuchs, D., Hausen, A., Wachter, H.: In: Biochemical and Clinical Aspects of Pteridines. First Workshop on Pteridines. St.Christoph, Arlberg (Austria), March 7 - 1o, 1982, pp 2o7-215, Eds.: H.Wachter, H.Ch.Curtius, W.Pfleiderer. Walter de Grüyter, Berlin - New York (1982)

21. Huber, Ch., Fuchs, D., Hausen, A., Margreiter, R., Reibnegger, G., Spielberger, M., Wachter, H.: In: Biochemical and Clinical Aspects of Pteridines. First Workshop on

Pteridines. St.Christoph, Arlberg (Austria), March 7-1o, 1982, pp. 227-232, Eds.: H.Wachter, H.Ch.Curtius, W. Pfleiderer. Walter de Gruyter, Berlin-New York (1982)

22. Margreiter, R., Fuchs, D., Hausen, A., Huber, Ch., Reibnegger, G., Spielberger, M., Wachter, H.: In: Biochemical and Clinical Aspects. First Workshop on Pteridines. St.Christoph, Arlberg (Austria), March 7-1o, 1982, pp. 233-24o, Eds.: H.Wachter, H.Ch.Curtius, W.Pfleiderer. Walter de Gruyter, Berlin-New York (1982)

URINARY NEOPTERIN EVALUATION IN MALIGNANT DISEASES IN CHILDHOOD

Christa Kostron-Krainz
Department of Pediatrics, University of Innsbruck

Dietmar Fuchs, Arno Hausen, Gilbert Reibnegger and Helmut Wachter
Institute of Medical Chemistry and Biochemistry, University of Innsbruck

Introduction

For many years clinicians are in search for a simple and reliable screening method for diagnosis and follow up in malignant diseases. Especially in children there is an increased demand for a method that does not frighten the patient too much. Urinary neopterin determination by high performance liquid chromatography has been shown to be of remarkable value in tumor diagnosis and monitoring therapy (1-3, 19,20). The method is rapid, sensitive and specific on one hand, on the other there is no additional encumbrance for the patient.
The present paper is designed to report preliminary results of a study on the behaviour of urinary neopterin in children with various malignancies : firstly, we intend to demonstrate the usefullness of neopterin as tumor marker also in the field of pediatrics. Secondly, the suitability of long term determinations of neopterin in urine of patients receiving therapy as a parameter for therapy control will be shown.

Patients and method

Urinary neopterin levels were measured on a group of 141 apparently healthy children of varying age in order to estab-

Biochemical and Clinical Aspects of Pteridines · Volume 1
Cancer · Immunology · Metabolic Diseases

lish age-dependent upper normal limits and on 50 children with different neoplasms, but free of viral infections.
The classification of the malignancies was as follows: 10 of the tumors were acute lymphoblastic leukemias (ALL), 4 acute myelocytic leukemias (AML), 5 neuroblastomas, 3 Wilms tumors, 3 Hodgkin's diseases, 4 non Hodgkin's lymphomas, 11 brain tumors and 10 rare malignant diseases, as well as 2 cases of systemic lupus erythematodes (SLE). 25 of the 50 patients were without chemotherapy or radiation therapy: either therapy had not started at the time of the investigation or treatment had been discontinued at least 6 months before.
For the treatment in patients with acute lymphoblastic leukemias we had different therapeutic regimens in different years (4-7), all with an intensive induction therapy, followed by maintenance therapy. The drugs given were prednisolone, vincristine, doxocrubicin, cytarabine, l-asparaginase, cyclophosphamide, 6-mercaptopurine, thioguanine as well as the intrathecally administered methotrexate. All patients had irradiation of cranial meninges for preventing meningeal leukemia.
For remission-induction phase in acute myelogenous leukemia we gave cytarabine, thioguanine, prednisolone, doxorubicin, followed by maintenance therapy with thioguanine and mercaptopurine and pulses of cytarabine, doxorubicin and prednisolone (8-10). Central nervous system prophylaxis was included. Therapy of patients with neuroblastoma, a malignant tumor originating from the symphathetic neuroblasts of the neural crest, depends on the age of the child, on the extent or stage of disease and the location of the primary tumor (11-13).
The therapeutic regimen of patients with Wilms tumor (nephroblastoma) is detailed elsewhere (13-15).
The treatment of patients with Hodgkin's disease and non Hodgkin's lymphoma is according to the internationally used treatment regimen (16). Details of therapy of rare malignant diseases are not mentioned here, as it would go too far.
Patients with systemic lupus erythematodes had an immunsuppressive therapy with prednisolone alone or in combination

with azathioprin.
Urine samples collected in the morning were employed for analysis; they were either analysed immediately or frozen at once and stored at -20°C in the dark to prevent oxidations or photochemical alterations in the composition. All analyses were performed without knowledge of the clinical diagnosis. To cancel out physiological variations in the concentration of the urine samples neopterin values were related to urinary creatinine measured simultaneously. The results were evaluated by means of comparison with the respective age group of the healthy control children: levels above the mean plus 2,5 standard deviations (S.D.) of each age group were considered elevated. Analyses were performed by the high performance liquid chromatography (HPLC) technique. The procedure of simultaneous detection of neopterin and creatinine in urine has been described in full detail elsewhere (17,18).

Results

Urinary neopterin in healthy children

Urine samples obtained from 149 apparently healthy children without detectable symptoms of viral infections were employed for defining age-dependent limits between normal and pathological ranges of neopterin excretion. Mean values and standard deviations are depicted in figure 1, table 1 also contains the upper normal limits (mean plus 2,5 standard deviations). Evidently urinary neopterin concentration is higher in the first years of life, decreasing gradually during growing age and eventually approximating the value determined for an adult control group (19,20) at the end of the second decade of life.

Fig.1 Urinary neopterin levels in healthy children

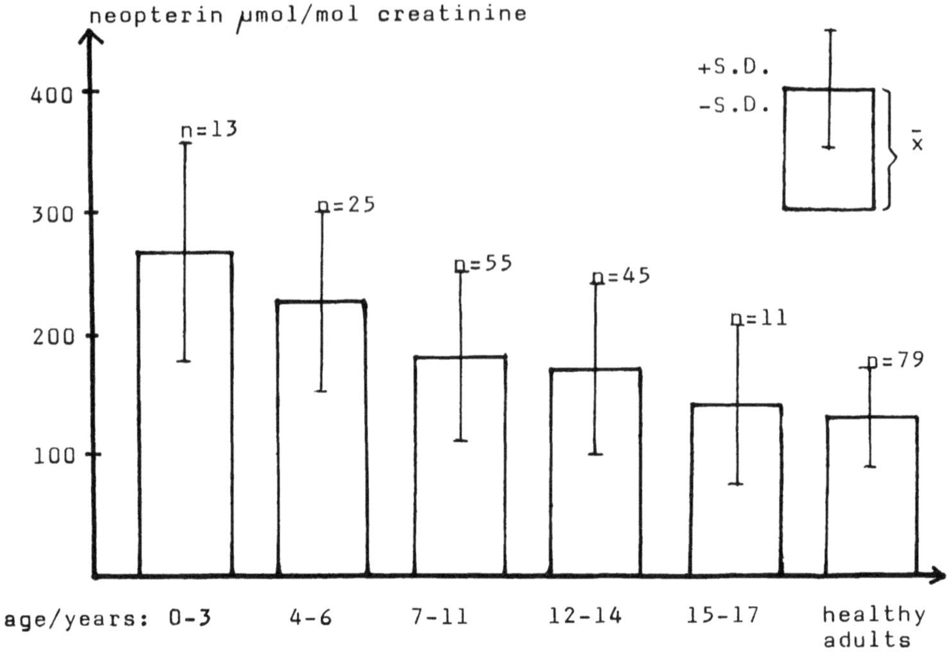

Table 1

Urinary neopterin levels in the control group

age/years	Number of children	$\bar{x}$	$\bar{x}$ + 2,5 SD
0 - 3	13	267	502
4 - 6	25	226	416
7 - 11	55	181	364
12 - 14	45	171	354
15 - 17	11	144	307
healthy adults	79	130	230

Urinary neopterin in children with various malignant neoplasms

The results of neopterin measurement in urine samples from 50 children with different malignancies are represented in detail in tables 2-9 (neopterin values are given in µmol/mol creatinine) including evaluation of correspondence between clinical findings and urinary neopterin in regard of the upper normal limits established in table 1. The results are summarized in table 10, indicating that in 41 (82 %) of the patients urinary neopterin is an accordance with diagnosis.

Legend to the tables 2-9 : CR complete remission
NR no remission
f + false positive
f _ false negative

Table 2

Urinary neopterin levels in patients with ALL

patient	age/years	clinical findings CR	NR	neopterin	correspondence yes	f+	f-
H.R.	13		+	356	+		
L.M.	5	+		247	+		
K.M.	2	+		999		+	
M.M.	4	+		250	+		
H.K.	11	+		206	+		
V.M.	8	+		309	+		
N.S.	12	+		242	+		
H.G.	4	+		292	+		
S.M.	10	+		65,5	+		
H.E.	10	+		115	+		

Table 3

Urinary neopterin levels in patients with AML

patient	age	clinical findings CR	NR	neopterin	correspondence yes	f+	f-
A.M.	3 weeks		+	2231	+		
V.R.	14 a	+		305	+		
I.V.	15 a	+		180	+		
K.K.	1 day		+	3147	+		

Table 4

Urinary neopterin levels in patients with non Hodgkin lymphoma

patient	age/years	clinical findings CR	NR	neopterin	correspondence yes	f+	f-
S.K.	9		+	2228	+		
K.M.	6	+		182	+		
B.S.	4		+	2022	+		
P.A.	9	+		314	+		

Table 5

Urinary neopterin levels in patients with Hodgkin's disease

patient	age/years	clinical findings CR	NR	neopterin	correspondence yes	f+	f-
L.M.	14	+		178	+		
F.K.	7		+	803	+		
V.A.	9	+		188	+		

Table 6

Urinary neopterin levels in patients with neuroblastoma

patient	age/years	clinical findings CR	NR	neopterin	correspondence yes	f+	f-
M.A.	3		+	549	+		
T.R.	3		+	1393	+		
Z.M.	12		+	303			+
E.A.	11	+		196	+		
W.P.	10/12	+		435	+		

Table 7

Urinary neopterin levels in patients with Wilms Tumor

patient	age/years	clinical findings CR	NR	neopterin	correspondence yes	f+	f-
M.S.	4	+		642		+	
P.A.	7	+		250	+		
V.R.	11	+		256	+		

Table 8

Urinary neopterin levels in patients with brain tumor

patient	age/years	clinical findings CR	NR	neopterin	correspondence yes	f+	f-
L.E.	5		+	823	+		
B.H.	15	+		255	+		
S.D.	2	+		1220		+	
R.P.	9	+		352	+		
L.A.	10		+	204			+
W.S.	9	+		112	+		
A.E.	12		+	664	+		
E.F.	12		+	148			+
G.E.	14		+	111			+
F.D.	7	+		71	+		
S.A.	3		+	271			+

Table 9

Urinary neopterin levels in patients with rare neoplastic diseases

patient	age (years)	clinical findings CR	NR	neopterin	correspondence yes	f+	f-
Rhabdomyo-Sa	2		+	304			+
s.Lupus e. 1	9	+		245	+		
s.Lupus e. 2	13		+	1941	+		
intraspinal T.	12		+	365	+		
Yolk sack T.	11	+		320	+		
Ewing Sa	12	+		52	+		
juv. CML	7		+	2580	+		
Dysgerminom	13	+		204	+		
Hystocytosis	12	+		103	+		
RMS	7	+		249	+		

Table 10

Correspondence between clinical diagnosis and urinary neopterin levels in children with malignant neoplasms without detectabel symptoms of viral infections

Type of malignancy	No. of patients	N. with correspondence
ALL	10	9
AML	4	4
NHL	4	4
Hodgkin's disease	3	3
neuroblastoma	5	4
brain tumor	11	6
Wilms tumor	3	2
rare malignancies	10	9
total	50	41 82 %

Figure 2

Urinary neopterin in follow up of a patient with NHL

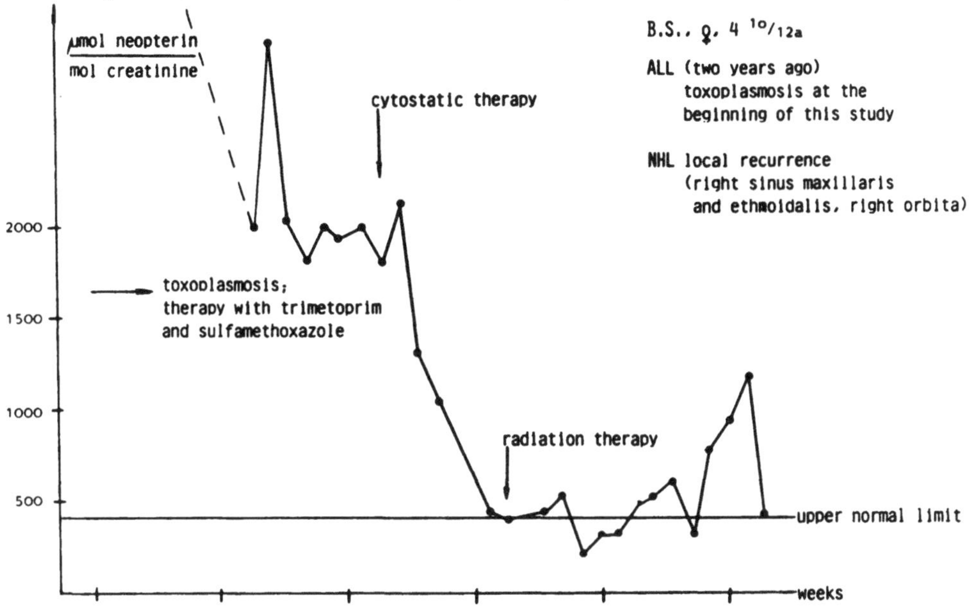

Figure 3

Urinary neopterin in follow up of a patient with jCML

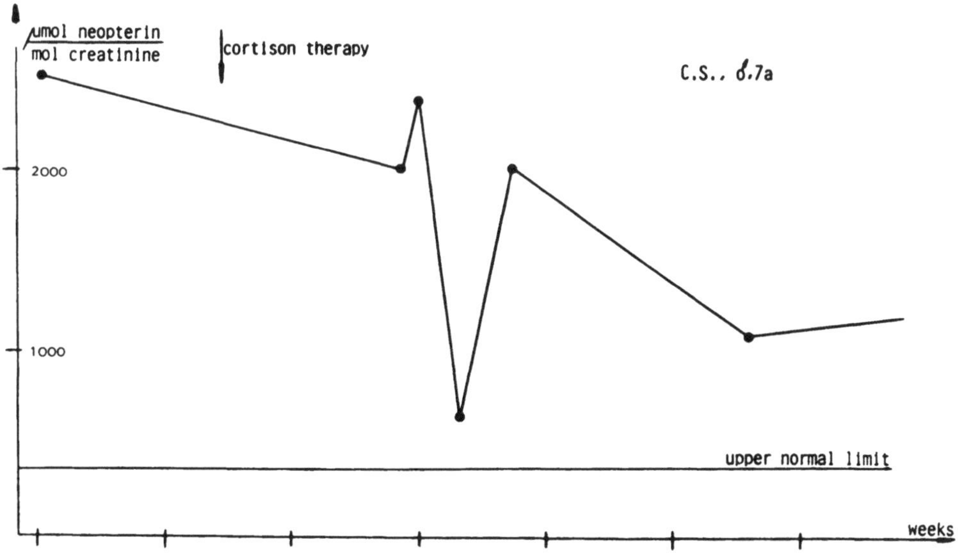

Urinary neopterin in long-term monitoring of therapy

Previous investigations have demonstrated the value of following therapy by long-term determinations of urinary neopterin (21, 22). Due to this fact we decided to start longitudinal studies also in the case of childhood cancer, preliminary results of which are disclosed in the present chapter.
The first patient is a 4 years old girl. 2 years ago an ALL had been diagnosed and treated. At the beginning of this study a toxoplasmosis was verified serologically by complement binding reaction and positive specific antibodies. An extraordinary high neopterin level in urine could be detected at that time, treatment by combined trimethoprim and sulfamethoxazole lowered neopterin concentration in urine, the level remained in the pathological range, however. Local recurrences of pathologic cells, found in the right sinus maxillaris and ethmoidalis and right orbita, - for this reason classified as Non Hodgkin's lymphoma (NHL) -, were treated by cytostatic and radiation therapy which was accompanied by decrease of urinary neopterin to only slightly elevated levels. Figure 2 represents the observed trend to neopterin excretion in urine.
The second patient, a boy of seven years suffering from chronic myelocytic leukemia of juvenile type, is represented in figure 3. Remission was never achieved and in correlation urinary neopterin always remained in the pathological range.
The urinary neopterin excretion of the next patient, a girl of 12 years with a glioblastoma multiforme, an extremely malignant brain tumor, is depicted in figure 4 again demonstrating the information content of this new parameter in monitoring therapy. During the course of radiation therapy neopterin excretion decreased transiently, but soon became elevated again, though the patient was still treated by radiation. A computed axial tomography (CT) did not reveal signs of tumor recurrence unequivocally, but a lumbar puncture, performed 2.5 weeks later, revealed cancerous cells

in cerebrospinal fluid, demonstrating that complete remission had not been achieved.

Figure 4

Urinary neopterin in follow up of a patient with glioblastoma multiforme

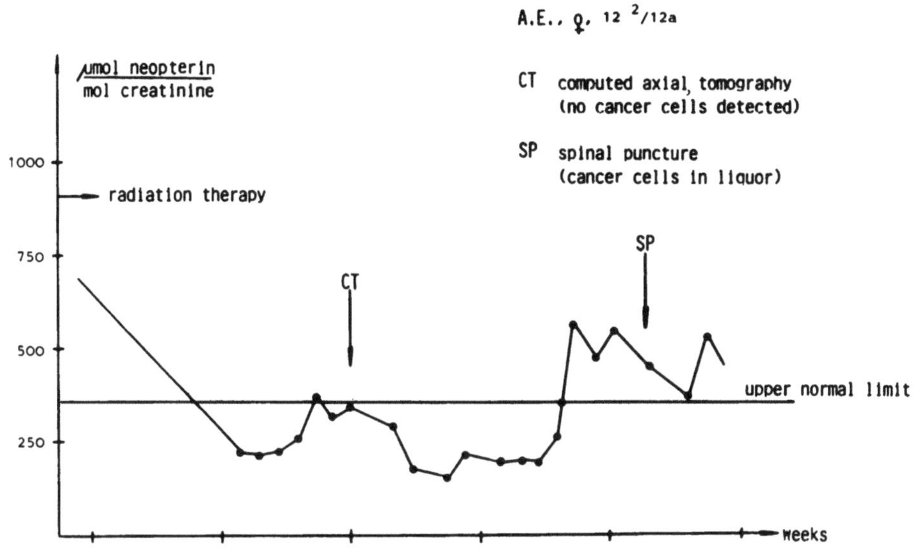

A very interesting feature could be observed in the follow up of a 13 years old girl with ALL, represented in figure 5. With few exceptions, neopterin in urine could be determined each day and be compared to the absolute number of lymphocytes of the respective day. Treatment by cytostatic drugs plus cortison resulted in a nearly parallel behaviour of urinary neopterin and the number of lymphocytes, the neopterin level being in the normal range. After cortison therapy was stopped we found a highly elevated neopterin excretion, suggesting rebound effects similar to results observed in allograft recipients (22). Further chemotherapy resulted in a marked lymphopenia, accompanied by highly elevated neopterin. Interruption of therapy was followed by an increase in lymphocytes and simultaneous decrease of urinary neopterin. When treatment was started again, lymphocytes began to decrease and, vice versa, urinary neopterin excretion started again to increase. These observations strongly suggests a mutual influence of neopterin metabolism and number of lymphocytes. The exact nature of this behaviour is not known, but perhaps there is some sort of regulation, where neopterin or some pteridines are involved.

Figure 5

Urinary neopterin and number of lymphocytes in a patient with acute lymphoblastic leukemia

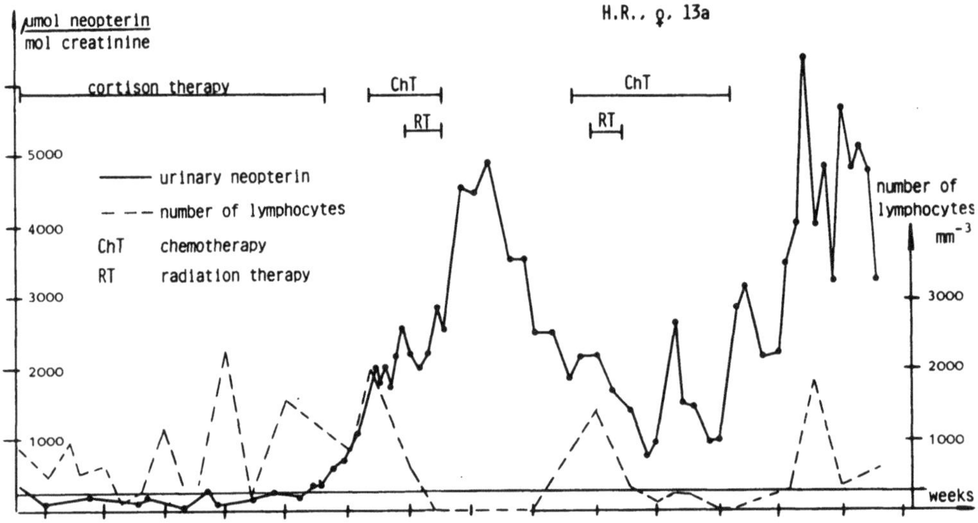

Discussion

As described previously, there is a high incidence of urinary neopterin elevation in active stage of malignant disease. Our preliminary results in the pediatric age group confirm these reports. We found significantly raised neopterin levels in patients with active malignancies, for this reason the determination of urinary neopterin seems to be a useful screening method. For diagnosis, the neopterin evaluation cannot replace the invasive methods as bon marrow aspiration, x-ray, computed axial tomographie or sonographie, but it will be a quick and economic supportive investigation. A group of 50 patients with malignant diseases comprised children with disorders of the hematopoetic system, the lymphoreticular system and with solid tumors and 2 patients with systemic lupus erythematodes. We found a correlation in 82 %. This high and significant incidence could not be obtained by any other biochemical method (carcinoembryonic antigen, α-Fetoprotein, cyclic Guanosinmonophosphate) in pediatric patients (23). Determination of neopterin is important in monitoring therapy. It allows an early indication of metastasis or progression of tumors. Collection of urine does not disturb children and can be performed in any age group. The analysis takes only a few minutes and can be done daily if necessary, which is finally essential for a useful clinical screening test. Up to this point in time the results are encouraging, further investigations are carried out emphatically. By now, we are busy to disclose the prognostic value of urinary neopterin in tumor diagnosis and to find out a possible relation between the neopterin values and tumor stage respectively the different groups of ALL.

The correlation between the absolute number of T-lymphocytes and neopterin-excretion during chemotherapy seems to be an important aspect in our opinion activated T-lymphocytes could be responsible for elevated urinary neopterin levels (19). In addition we investigated neopterin-excretion in patients with inborn immun deficiency to get further information about the biological background of increased neopterin production.

References

1. Wachter, H., Hausen, A., Graßmayr, K.: Erhöhte Ausscheidung von Neopterin im Harn von Patienten mit malignen Tumoren und mit Viruserkrankungen. Hoppe-Seyler's Z.Physiol. Chem. 360: 1957-1960 (1979).

2. Rokos, H., Rokos, K., Frisius, H., Kirstaedter, H.J.: Altered urinary excretion of pteridines in neoplastic disease. Determination of Biopterin, Neopterin, Xanthopterin and Pterin. Clin.Chim.Acta 105: 275-286 (1980).

3. Stea, B., Halpern, R.M., Halpern, B.C., Smith, R.A.: Urinary excretion levels of unconjugated pterins in cancer patients and normal individuals. Clin.Chim.Acta 113: 231-242 (1981).

4. Pinkel, D., Hernandez, K., Borella, L. et al.: Drug dosage and remission duration in childhood lymphocytic leukemia. Cancer, 27: 247-256 (1971).

5. Pinkel, D.: Treatment of acute leukemia. Pediatr. Clin. North Am., 23: 117 (1976).

6. Riehm, H., Gadner, H., Welte, K.: Die West-Berliner Studie zur Behandlung der akuten lymphoblastischen Leukämie des Kindes - Erfahrungsbericht nach 6 Jahren. Klin. Pädiatr. 189: 89-102 (1977).

7. Henze, G., Langermann, H.-S., Brämswig, J., Breu, H., Gadner, H., Schellong, G., Welte, K., Riehm, H.: Ergebnisse der Studie BFM 76/79 zur Behandlung der akuten lymphoblastischen Leukämie bei Kindern und Jugendlichen. Klin. Pädiatr. 193: 145-154 (1981).

8. Weinstein, H.J., et al.: Treatment of acute myelogenous leukemia in children and adults. N.Engl. J.Med. 303: 473-8 (1980).

9. Scheer, U., Schellong, G., Riehm, H.: Verbesserte Prognose der akuten myeloischen Leukämien bei Kindern nach intensivierter Anfangstherapie. Klin. Pädiatr. 191: 210-216 (1979).

10. Plüss, H.J., Hitzig, W.H.: Die akuten myeloischen Leukämien im Kindesalter. Schweiz.med.Wschr. 110: 1459-1462 (1980).

11. Jentsch, K.: Die Strahlentherapie des Neuroblastoms. Wien. klin.Wschr. 89, Suppl. 67 (1977).

12. Kurz, R., Mutz, I., Pichler, E.: Neuroblastom: Diagnose, Therapie und Prognose an Hand von 56 Patienten. Wien.klin. Wschr. 89, Suppl. 67 (1977).

13. Howanitz, L., Hubmer, G., Sauer, H.: Chirurgisches Vorgehen bei Wilms-Tumoren und Neuroblastomen. Wien.klin. Wschr. 89, Suppl. (1977).

14. Kurz, R., Pichler, E., Mutz I.: Wilms-Tumoren: Diagnose, Therapie und Prognose an Hand von 43 Patienten. Wien.klin. Wschr. 89, Suppl. 67 (1977).

15. Jentzsch, K.: Sie Strahlentherapie des Nephroblastoms. Wien. klin.Wschr. 89, Suppl. 67 (1977).

16. Jenkin, R.D.T., Morris Jones, P.: Cancer in children. Berlin-Heidelberg-New York: Springer (1975).

17. Hausen, A., Fuchs, D., König, K., Wachter, H.: Determination of neopterin in human urine by reversed phase-high-performance liquid chromatography. J.Chromatogr., 227; 61 (1982).

18. Hausen, A., Fuchs, D., Wachter, H.: Determination of creatinine in urine by HPLC. J.Clin.Chem.Clin.Biochem. 16: 375-378 (1981).

19. Fuchs, D., Hausen, A., Huber, C., Margreiter, R., Reibnegger, G., Spielberger, M., Wachter, H.: Hoppe-Seyler's Z.Physiol.Chem. 363: 661-664 (1982).

20. Hausen, A., Fuchs, D., Grünewald, K., Huber, H., König, K., Wachter, H.: Urinary neopterine in the assessment of lymphoid and myeloid neoplasia, and neopterine levels in haemolytic anaemia and benign monoclonal gammopathy. Clin. Biochem. 15 (1): 34-37 (1982).

21. Bichler, A., Fuchs, D., Hausen, A., Hetzel, H., König, K., Wachter, H.: Urinary neopterine excretion in patients with genital cancer. Clin. Biochem. 15 (1): 38 - 40 (1982).

22. Hausen, A., Fuchs, D., Grünewald, K., Huber, H., König, K., Wachter, H.: Urinary neopterine as marker for haematological neoplasias. Clin.Chim.Acta 117: 297-305 (1981).

23. Puschendorf, P., Grunicke, H., Jentsch, J., Berger, H.: Urinary excretion of cyclic guanosine 3',5'-monophosphate in children with malignant tumours. J.Clin.Chem.Clin.Biochem. 19: 979-981 (1981).

NEOPTERIN ESTIMATIONS IN BENIGN AND MALIGNANT HEMATOLOGICAL DISORDERS AND THEIR CONNECTIONS WITH TUMOR STAGING AND REMISSIONS

Kurt Grünewald, Heinz Huber

Department of Internal Medicine, University of Innsbruck

Dietmar Fuchs, Arno Hausen, Gilbert Reibnegger, Helmut Wachter

Department of Medical Chemistry and Biochemistry,
University of Innsbruck

Introduction

There are many natural sources of unconjugated pteridines (the term "pteridine" is used to refer to the bicyclic heterocyclic pyrimido-(4,5b)-pyrazine; the 2-amino-4-hydroxy-pteridine is designated pterin). However, the only well established biological function of pteridines is that of tetrahydrobiopterin which serves as a cofactor in the enzymatic hydroxylation of phenylalanine, tyrosine and tryptophan by mixed function monooxygenases (1). Pteridines are synthesized from guanosine triphosphate (GTP) in a reaction which is catalysed by a cyclohydrolase and which involves the removal of the 8-carbon followed by condensation with the 1- and 2-carbons of ribose to form a pyrazine ring. The first isolated product is dihydroneopterin-triphosphate, the precursor for all natural pteridines. Recently there has been evidence that pteridine excretion is indicative of cell proliferation (2). High urinary pteridine levels have been reported both in laboratory animals (3) and in patients with various malignancies (4,5) and neopterin has been identified as an excretion product of activated T-lymphocytes (6). Thus the role of neopterin in tumor immunology has stimulated considerable interest.

Biochemical and Clinical Aspects of Pteridines · Volume 1
Cancer · Immunology · Metabolic Diseases

The choice of hematological neoplasias for study here was influenced by the following considerations:

1) Most hematological neoplasias are well defined disorders.
2) There exists a lot of prognostic parameters with which to correlate neopterin levels.
3) These malignancies represent a wide spectrum of cell differention and kinetics.
4) Most hematological neoplasias can be accurately staged.

Patients and Methods

Urinary neopterin levels were measured in 193 patients with hematological neoplasias (52 in remission). Urine of the control group was collected from 208 male and from 209 female volunteers. The neopterin values were related to the urinary creatinine concentration. In each urinary sample neopterin was estimated simultaneously with creatinine by HPLC on reversed phase. 1.0 ml aliquots of urine samples were pressed through SEP-PAK C_{18} cartridges and eluted with 9 ml water. A 10-μl aliquot of the mixed elutions was injected into the column. High pressure liquid chromatography was performed on a Varian Model LC 5000 instrument equipped with an automatic sample injection-system. Prepacked columns (300 x 3.9 mm, 10 μ-Bondapack C_{18}) were used. Chromatographic elution was performed with an aqueous 15 mmol/l potassium phosphate buffer (pH = 6.4). For the first 7 minutes the flow rate was 0.8 ml/min. Using a flow (0.8 - 1.4 ml/min) and composition gradient (0 - 1.218 mol/l acetonitrile) a total analysis time of 19 minutes was reached. The effluent was monitored by setting the fluorescence detector (Model SFM 23, Fa. Kontron) for an excitation wavelength of 353 nm and an emission of 438 nm. Chromatograms were quantified by means of a chromatography data system (Model CDS 111, Varian).

Results

Urinary neopterin in healthy adults

Urinary samples (n = 417), obtained from 208 male and 209 female individuals were used to assess the normal range. Table 1 shows the mean and the median urinary neopterin levels for different age groups.

Table 1

Urinary neopterin levels (µmol neopterin/mol creatinine) in healthy individuals (n = 417) for different sex and age groups.

Age years	sex	no of subjects	$\bar{x} \pm SD$	limit of tolerance+)	interquartile range	median
18-25	♂	42	123 ± 30	195	102-151	120
26-35	♂	29	101 ± 33	182	71-121	87
36-45	♂	41	109 ± 28	176	87-132	108
46-55	♂	32	105 ± 36	197	73-136	98
56-65	♂	31	119 ± 39	218	81-157	120
>65	♂	33	133 ± 38	229	106-157	134
18-25	♀	55	128 ± 33	208	103-155	124
26-35	♀	28	124 ± 33	209	95-150	129
36-45	♀	31	140 ± 39	239	113-168	134
46-55	♀	28	147 ± 32	229	126-165	141
56-65	♀	26	156 ± 35	249	131-178	158
>65	♀	41	151 ± 40	251	110-184	143
all healthy males		208	115 ± 36		87-139	112
all healthy females		209	140 ± 40		120-186	147

+) limit of tolerance was tested according to Lieberman (7).

The values for healthy men of all ages were significantly lower than those for women ($p < 0.01$). There was also a correlation between the neopterin/creatinine ratio and age, which was more noticeable in males. This dependence may be mainly due to urinary creatinine levels varying distinctly with sex and age. To test the utility of the neopterin assay, upper confidence limits for each group of healthy subjects were calculated. These limits, shown in Table 1, include with 95% probability 97.5% of healthy subjects. These correspond to about $\bar{x}$ + 3SD.

Hodgkin's disease

Definition

Hodgkin's disease differs in many respects from most neoplastic disorders. In particular there is no clear cut homogenous tumor cell population and infiltration of non-lymphatic tissues occurs late in the course of the disease. Based on the evaluation of DNA-synthesizing lymphocytes and an increased turnover after $[^3H]$ thymidine labelling in vivo, Ch. Huber and G. Michlmayr (8,9) suggested that there was an increased renewal of blood lymphocytes in patients with Hodgkin's disease. The majority of these DNA-synthesizing cells were of the T-cell series. These and other immunological findings indicate that T-cells may be infected by a tumor-inducing virus which causes a change in surface antigenity such that normal T-cells react against these altered cells in a graft-versus-host type reaction. This may lead to the production of a malignant reticulum cell line. Shiftan (10) proposed the hypothesis that Hodgkin's disease could result from a viral infection of a lymph node cell leading to a further immune response with a defective effector arm. It is interesting that neopterin levels are strongly elevated in viral diseases and Fuchs et al. (6) showed that neopterin is also a marker for the early prediction of human allograft rejection, a response of T-cells. The results of treatment of patients with Hodgkin's disease continue to improve. A disease that was considered uniformly fatal only

a few decades ago, can now be cured in the majority of newly diagnosed patients. This changing outlook was made by improved combined modality therapy (high energy irradiation in stages I, II and III A and of cyclic intensive combination chemotherapy in patients with advanced nodal and extranodal involvement) and staging which include: physical examination, laparotomy, lymph nodes and bone marrow biopsies, chest x-ray and computer tomography. Intensive cyclic therapy with mechloretamine, vincristine, procarbazine and prednisone (MOPP) results in complete remission in 70-80% of patients (11). Prolonged disease-free survival in MOPP resistant Hodgkin's disease has been found after treatment with adriamycin, bleomycin, vinblastine and dacarbacine (12). It is sometimes paradoxical that this dramatic improvement in survival has made the choice of optimum management more difficult because long observation of large groups of patients is needed to distinguish between the results of available treatment options and this prolonged survival has led to the recognition of an increased risk of second tumor induction. Hence the benefits of this approach in Hodgkin's disease must be weighed against its carcinogenic potential (13,14). Four subgroups of Morbus Hodgkin have been recognized: Lymphocyte predominance, nodular sclerosis, mixed cellularity, lymphocyte depletion. Table 2 shows the staging classification of Hodgkin's disease, known as the Ann Arbor classification. The absence or presence of fever, night sweats or unexplained loss of 10% or more of body weight in the 6 months before admission are denoted by the suffix letters A & B, respectively. The stage or extent of disease at the time of diagnosis is the most important single guide to prognosis and treatment. Extension below the diaphragm is a rare event in the asymptomatic patient with a negative staging laparotomy. The treatment of symptomatic subjects, particularly those with disease above and below the diaphragm, is less satisfying.
Results are shown in Table 3.

Table 2

Staging classification of Hodgkin's disease, modified from Carbone et al. (15).

Stage	Definition
I	Involvement of a single lymph-node region
II	Involvement of two or more lymph-node regions on the same side of the diaphragm (II), which may also be accompanied by localized involvement of an extralymphatic organ or site (II_E).
III	Involvement of lymph-node regions on both sides of the diaphragm (III), which may also be accompanied by involvement of the spleen (III_S) or by localized involvement of an extralymphatic organ or site (III_E).
IV	Diffuse or disseminated involvement of one or more extralymphatic organs or tissues, with or without associated lymph-node involvement.

Table 3

Neopterin levels in patients with Morbus Hodgkin according to stage and remission.

patients	n	neopterin (μmol) / creatinine (mol)		% in the pathological range
stage I and II	7	$\bar{x}$	321	
		S.D.	50	71
		C.V.	15	
stage III and IV	9	$\bar{x}$	367	
		S.D.	164	100
		C.V.	45	
all patients stage I - IV	16	$\bar{x}$	347	
		S.D.	130	88
		C.V.	37	
remission	24	$\bar{x}$	231	
		S.D.	96	21
		C.V.	42	

Non-Hodgkin Lymphomas

Definition

The malignant non-Hodgkin Lymphomas (NHL) constitute a variety of lymphoproliferative disorders that often present as clinically advanced disease. The controversy over the nature of NHL is reflected in the existence of multiple classification. For our patients we have chosen the Kiel classification which is shown in Table 4.

Table 4
Kiel classification (16,17)

LOW-GRADE MALIGNANCY	
I lymphocytic:	B-CLL T-CLL hairy cell leukemia mycosis fungoides and Sezary syndrome T-zone
II lymphoplasmacytoid (immunocytic)	
III centrocytic	
IV centroblastic-centrocytic (follicular and/or diffuse)	
HIGH-GRADE MALIGNANCY	
I centroblastic	
II lymphoblastic (B, T-cell type, unclassified including null type)	
III immunoblastic	

Up to now many NHL have been identified as T- and B-cell derived lymphoid tumors. In the group of low grade lymphomas B-cell disorders predominate, whereas lymphomas of the T-lymphocytes are commonly found in patients with lymphoid tumors of high malignancy (18). Staging of NHL is the same as in Morbus Hodgkin. Ann Arbor staging classification has unfortunately not the same prognostic value as in Morbus Hodgkin so the prognostic relevance of histological subgroups, cell marker studies, bone marrow function, age of patient, sites of involvement, the "tempo" of tumor progression, the number of DNA-synthesizing cells and clinical laboratory parameters are often more successful (19,20,21,22). Silvestrini (23) found a correlation of low $\left[{}^{3}H\right]$ thymidine labelling index with low grade malign lymphomas. Although the DNA-synthesis time was found to be within the normal range, the generation time of these neoplastic lymphocytes seemed to be prolonged. Even with a normal growth fraction, the total number of mitotically active neoplastic cells is increased in each NHL due to extensive enlargement of the total cellularity of each patient's lymphatic system. The total number of newly produced lymphoma cells is significantly increased. The NHL of low-grade malignancy are considered to be characterized by the proliferative constellation of an enhanced cell production associated with reduced or normal cell death rates, whereas in the NHL of high malignancy increased cell death rates seem to be overcompensated by even higher cell renewal rates (24). NHL are highly responsive to therapy (20). Influence of chemotherapy is hard to estimate in lowgrade malignancy NHL because about 70% of non-treated patients are alive 6 years after diagnosis. High-grade malignancy NHL show about 50% complete remissions with a mean duration between 2 and 3 years (25). While in histological types with a good prognosis a mild regimen may suffice, aggressive cytotoxic regimens are preferred in treating those with a less favourable prognosis. As in Morbus Hodgkin in stages I and II (localized regional stages) radiotherapy alone is sufficient; in advanced stages combined cytotoxic regimens are the accepted form of treatment.

The following different chemotherapy regimens were used:

Table 5
Therapy of non-Hodgkin lymphomas

Low-grade malignancy NHL:

1)	Chlorambucil/prednisone (like CLL)		
2)	COP:	cyclophosphamide	750 mg/m^2 i.v.+) day 1+8
		vincristine	1.5 mg i.v. day 1+8
		prednisone	40 mg p.o.++) day 1-14
		repeated every month	

High-grade malignancy NHL:

1)	CHOP:	cyclophosphamide	750 mg/m^2 i.v. day 1
		adriblastin	50 mg/m^2 i.v. day 1
		vincristine	1.5 mg i.v. day 1
		prednisone	100 mg/ daily x 5 p.o.
		course repeated every 2-3 weeks	
2)	BACOP:	adriblastin	25 mg/m^2 i.v. day 1+8
		cyclophosphamide	650 mg/m^2 i.v. day 1+8
		vincristine	1.5 mg/m^2 i.v. day 1+8
		bleomycin	5 mg/m^2 i.v. day 15+21
		prednisone	60 mg/m^2 p.o. day 15-29
		repeated every 4 weeks	

+) intravenously, ++) orally

Table 6
Neopterin levels in patients with non-Hodgkin lymphomas according to stage and remission.

patients	n	neopterin (µmol) / creatinine (mol)		% in the pathological range
stage I and II	14	$\bar{x}$	497	79
		S.D.	256	
		C.V.	52	
stage III and IV	19	$\bar{x}$	915	100
		S.D.	471	
		C.V.	52	
stage I - IV	33	$\bar{x}$	710	91
		S.D.	440	
		C.V.	62	
remissions	13	$\bar{x}$	193	8
		S.D.	70	
		C.V.	36	

Chronic Myelocytic Leukemia (CML)

Definition

CML belongs to the chronic myeloproliferative disorders. The cause of the malignant transformation of hematopoietic cells in human chronic myelocytic leukemia is unknown. Studies of the Philadelphia Chromosome and the x-linked enzyme glucose-6-phosphate dehydrogenase have implied that CML originates from a monoclonal stem cell (26,27). Involvement of the B-lymphoid system was detected in the last few years (28). Diagnoses: Leukocytes count $>40,000/mm^3$, granulocytes count $>80\%$ of total leukocytes, $<30\%$ myeloblasts and promyelocytes, no hiatus leukemicus, splenomegaly, leukocyte alkaline phosphatase lowered or zero, vitamin B_{12} serum levels 3-10 x higher than normal, Philadelphia Chromosome in 80-90% (29). Estimation of $[^3H]$ thymidine labelling index often shows indices of 10-20% in the early chronic stages which is high in comparison to the blast crisis. There is therefore a good negative correlation between labelling index and the proportion of myeloblasts in the marrow (30).Reported medium survivals range from 21 to 43 months (31,32). Improved accuracy in prognosis at time of diagnosis would be of help. There have been surprisingly few prognostic factors (32,33) indentified in CML: females live longer than males and patients having the Philadelphia Chromosome live longer than those without. Thrombocytopenia is an unfavourable sign. Shilling et al. (33) did not find that the proportion of blasts and promyelocytes in the peripheral blood at time of diagnosis correlated with survival. Collins R.K. et al. (34) found that small splenic size in CML is associated with longer survival. The common terminal event in CML is transformation into acute leukemia. It appears that the features observed at the time of diagnosis have little value in predicting the time of onset of blastic transformation. There is no clear improvement of survival by chemotherapy. We started treatment when performance status was strongly lowered by forthcomming anemia, thrombocytopenia or thrombocytosis, progressive enlargement of spleen and doubling time of leukocytes was shorter than 30 days. The most common chemotherapeutic agent was busulfan (4-6 mg/day) until the leukocytes dropped to 15,000/mm^3. Then we stopped the therapy as long as no progress in disease was seen. In cases where the leukocyte count did not drop to 15,000 or splenomegaly was not decreased, busulfan (1-2 mg/day) was given continously.

Table 7

Neopterin levels in 21 patients with CML (95% showed levels above the upper normal limit)

	µmol neopterin / mol creatinine	Leuko ($x10^3/mm^3$)	Thrombo ($x10^3/mm^3$)	Hb (g/dl)	LDH (U/l)
$\bar{x}$	525.2	33.9	464.5	12.8	337.7
S.D.	265.9	39.0	196.3	1.92	174.0
C.V.%	50.6	115.2	42.3	15.0	51.5

µmol neopterin / mol creatinine	$\bar{x}$	S.D.	C.V.%
patients with LDH[1)] more than 250 U/l:	583	317	54.4
patients with LDH less than 250 U/l:	482	224	46.4
patients with splenomegaly[2)]	720	273	38
patients without splenomegaly[3)]	371	130	35

1) serum lactic dehydrogenase (E.C.1.1.1.27)

2) white blood cells: $\bar{x}$ = 52 ($x10^3/mm^3$)
platelets: $\bar{x}$ = 426

3) white blood cells: $\bar{x}$ = 20
platelets: $\bar{x}$ = 401

In all CML patients we investigated, common prognostic parameters have been compared with neopterin levels and we are starting a follow-up-study to clearify, whether neopterin levels would be an early parameter for the onset of blast crisis. In this respect it is noteworthy, that Schöch et al. (35) found highly elevated total excretion of two normal (adenine, guanine) and four modified nucleobases (methylated guanine derivates) with high performance liquid chromatography correlated with the number of leukocytes in a follow-up-study of CML. Results were shown in Table 7.

Chronic lymphocytic leukemia (CLL)

Definition

CLL has been considered as a cumulative disease of immunological incompetent but long-lived lymphocytes (36). Lymphocytic infiltration is found in blood and bone marrow, lymph nodes, spleen, liver and sometimes in salivary gland, epidermis and lung. Diagnosis can be made if there are more than 15,000 lymphocytes/mm^3 in blood and lymphocytes should compare more than 30% of nucleated bone marrow cells. Immunological investigations provided evidence that CLL is an irreversible, monoclonal proliferation (37) of B-lymphocytes. T-lymphocytes are relatively lowered in number but their function is usually not disturbed (38). A defect of maturation of the leukemic clone into immunoglobulin secreting cells which have a short half life, along with a defect in the recirculation of the lymphocytes, could account for the accumulation of leukemic cells (39). The proliferating cell fraction in CLL is quite small (40,41). Stryckmans (42) and Moayeri (43) found higher in vitro uptake of tritiated $[^3H]$ thymidine by leukocytes of a standard volume of peripheral blood, that was associated with a higher lymphocyte count, a more advanced stage, greater frequency of functional impairment and shorter survival. Van der Woerd de

Lange et al. (44) found abnormal numbers of proliferating T-lymphoid cells and an increased number of proliferating non-T-lymphoid cells in clinical stages O - I. Stages III - IV demonstrate a significant increase of the proliferation rate of both T and non-T-lymphoid cells. Prognosis in CLL varies considerably; because the life span of leukemic lymphocytes in CLL is very long, high white cell count does not necessarily imply very active disease. A real progress has been made since Rai et al. (45) introduced a clinical staging system well correlated to prognosis. This staging system is shown according to median survival time in Table 8.

Table 8 (29)
Staging system of CLL according to median survival time

Stage	Physical examination and laboratory data	median survival time (months)
O	lymphocytosis>15,000/mm^3	150
I	lymphocytosis plus enlarged lymph nodes	101
II	lymphocytosis plus enlarged liver or spleen	71
III	lymphocytosis plus anemia	19
IV	lymphocytosis plus thrombocytopenia	19

We usually started treatment in stages III and IV but also in stages I and II, if there was a clear progression of splenomegaly or lymph-node involvement or if complications developed resulting from the disease. The most common treatment was according to Knospe et al. (46): Chlorambucil 10 mg/day on days 1-3 orally plus prednisone 60 mg on day 1, 40 mg on day 2 and 20 mg on day 3. This therapy was maintained every 2 weeks elevating the chlorambucil dose by 5 mg until 20 mg/day on days 1-3 was reached. Therapy was stopped when leukocyte counts were lower than 10,000/mm^3. In some cases COP regimen (cyclophosphamide, vincristine, prednisone) or total body irradiation (total dose 0.6-3.6 Gy) have been used. We examined neopterin levels according to stage and response to therapy; attention should be paid to the connection between neopterin, nucleotide-metabolizing enzymes and lymphocyte differentiation for a better understanding of the biochemical and pathophysiological implications of neopterin excretion (47).

Table 9
Neopterin levels in patients with CLL

patients	n	neopterin (μmol) / creatinine (mol)		% in the pathological range
all patients	26	$\bar{x}$	593	96
		S.D.	273	
		C.V.	46	
stage I and II	12	$\bar{x}$	386	92
		S.D.	151	
		C.V.	39	
stage III and IV	14	$\bar{x}$	771	100
		S.D.	224	
		C.V.	29	
without hepato-splenomegaly	10	$\bar{x}$	355	92
		S.D.	133	
		C.V.	37	
with hepato-splenomegaly	16	$\bar{x}$	742	100
		S.D.	227	
		C.V.	31	
patients with LDH $>$ 250 U/l : $\bar{x}$ 725 S.D. 229 C.V.% 31.6				
patients with LDH $<$ 250 U/l : $\bar{x}$ 504 S.D. 281 C.V.% 55.7				

Multiple Myeloma (M.M.)

Definition

Multiple Myeloma is a monoclonal proliferation of immunoglobulin secreting cells. It is characterized by an infiltration of the bone marrow caused by neoplastic plasma cells which in most instances produce a myeloma protein and/or one of its constituent polypeptide chains. A number of factors such as renal function, hemoglobin, serum calcium, release of urinary Bence Jones protein, immunoglobulin type, extent and type of bone lesions, serum albumin and performance status, at the onset of disease are known to be important in relation to survival (48). Monoclonal immunoglobulin production predicts stage with reasonable accuracy, but as a single parameter it does not predict survival. This abnormality is probably due to the fact that a high M-component (monoclonal immunoglobulin) level can be due to either a small number of cells producing large quantities of M-component per cell or a large number of cells producing only a small amount of M-component per cell. Therefore other parameters are often useful. On the basis of the mathematical relationship (49):

$$\text{Total body myeloma cell number} = \frac{\text{total body monoclonal immunoglobulin}}{\text{cellular monoclonal immunoglobulin synthetic rate}}$$

a clinical staging system was developed by B.G.M. Durie and S.E. Salmon (48). Use of this clinical staging system provides better initial assessment and follow up of individual patients. Even this subdivision of patients into stages I, II and III (low, intermediate and high tumor cell mass) plus stratification (A or B) for serum creatinine $\geq$ 2.0 mg/100 ml has not entirely accounted for all prognostic differences. It therefore seemed that the kinetics of tumor growth and regression might have prognostic importance. A $[^3H]$ thymidine labelling index of $< 1\%$ was associated with long survival in each patient group. A pretreatment LI of $> 3\%$ in high cell mass patients with a high total number of DNA-synthesizing cells conferred a very poor prognosis (50).

Radiation therapy was the only effective treatment until a number of chemotherapeutic agents were also found to effective.

Table 10

Myeloma Staging System (48)

Stage	Criteria	Measured myeloma cell mass (cells x $10^{12}/m^2$) +)
I	All of the following: 1) Hemoglobin value $>$ 10 g/100 ml 2) Serum calcium value normal ($\leq$12 mg/100 ml) 3) On roentgenogram: normal bone structure (scale 0) or solitary bone plasmacytoma only 4) Low M-component production rates a) IgG value $<$ 5 g/100 ml b) IgA value $<$ 3 g/100 ml c) Urine light chain M-component on electrophoresis $<$ 4 g/24 hours	$<$0.6 (Low)
II	Fitting neither Stage I nor Stage III	0.6-1.20 (Intermediate)
III	One or more of the following: 1) Hemoglobin value $<$8.5 g/100 ml 2) Serum calcium value $>$ 12 mg/100 ml 3) Advanced lytic bone lesions (scale 3) 4) High M-component production rates a) IgG value $>$7 g/100 ml b) IgA value $>$5 g/100 ml c) Urine light chain M-component on electrophoresis $>$ 12 g/24 hours	$>$1.20 (High)

Subclassification

A = Relatively normal renal function (serum creatinine value $<$ 2.0 mg/100 ml)

B = Abnormal renal function (serum creatinine value $\geq$ 2.0 mg/100 ml)

Examples

Stage IA = low cell mass with normal renal function
Stage IIIB = high cell mass with abnormal renal function

+) 10^{12} cells = approximately 1 kg; m^2 = square meter of body surface area.

Unfortunately, these agents are not curative and radiation therapy continues to be used as an adjuvant for patients with disseminated disease. Solitary plasma cell lesions of bone or soft tissue are uncommon - They are usually treated with local field radiation therapy with or without chemotherapy. Low doses of radiation are usually adequate to treat painful bone lesions (15-20 Gy) of multiple myeloma and doses of 50-65 Gy in six to seven weeks are recommended for solitary plasmacytomas (51).

The addition of vincristine to different chemotherapy regimens including adriamycin or alkylating agents combined - whith shortening the treatment interval - was associated with higher response rates and with a longer survival time (52). The most common therapy we used was: Melphalane 10 mg/day plus prednisone 40 mg/day for 7 days. Other protocols included cyclophosphamide and vincristine or, in particularly resistant patients adriamycin and nitrosureas. We treated patients in stage II and III of the disease. In stage I only those with definite progressive disease were exposed to cytotoxic therapy. The mean survival time of patients with multiple myeloma was about 3 years (52) and causes of death were: progressive myeloma, renal failure, sepsis and bleedings (52,53). Quantification of serum paraprotein with other staging parameters gives us relative good information about tumor mass and response on treatment. Therefore neopterin levels in these patients were investigated for possible correlation with these approved parameters (Table 11). Further investigations, especially follow-up-studies, are still preliminary but successful results may be expected.

Benign monoclonal gammopathy

Definition

Individuals are asymptomatic and without evidence of plasma cell proliferation in the marrow. Anemia, bone lesions and renal disease are absent. The monoclonal immunoglobulin in the serum does not change significantly over the years and only a few patients later develop multiple myeloma. Patients with benign gammopathy have excellent survival and plasma cells have only a low [^{3}H]thymidine labelling index in comparison with multiple myeloma (50). It seems likely that most of these individuals have a benign disorder, perhaps as a pronounced immune response to some unknown antigen and/or defects of immune regulation. In this context it is of interest that individuals with deficient T-cells often show benign paraproteinaemia and that athymic mice have a high incidence too. Regardless of the ultimate nature of this disorder, chemotherapy is neither useful nor necessary. Results are shown in Table 11.

Table 11

Neopterin levels in patients with multiple myeloma and benign paraproteinaemia.

	M.M. Stage II and III	M.M. Stage I	Benign Paraproteinaemia
n	18	9	9
neopterin $\bar{x}$	385	249	167
S.D.	158	188	58
C.V.%	41	75	35
range	187-771	123-738	135-206
median	408	208	146
% in the pathological range	70	29	0

Response to therapy with clinical remissions in M.M.:

n = 7; neopterin $\bar{x}$: 147 S.D. : 36 C.V.% : 24%; range: 118-219 median: 139, no patient in the pathological range.

Acute Leukemia

Definition

Leukemia is a disease characterized by neoplastic proliferation of one of the blood forming cells. Because there are significant differences in their age distribution and responsiveness to treatment, the acute leukemias are commonly divided into two major types: acute lymphoblastic leukemia (ALL) and acute non-lymphoblastic leukemia. The latter is further divided into several subtypes: myeloblastic (AML), myelomonocytic (AMML), monoblastic or monocytic (AMOL), promyelocytic (PML), erythroleukemia (AEL) and several other rare types. A fundamental abnormality is a failure of cell differentiation with resultant accumulation of immature leukocytes or lymphocytes. This accumulation of blast cells in the blood and bone marrow is combined with the failure of normal hemopoiesis. The leukemogenesis occurs at the level of a pluripotent hemopoietic stem cell. Although leukemic blast cells proliferate more slowly than normal precursors, they accumulate because maturation, which leads quickly to cell death, does not proceed normally (54).
A number of articles have been published regarding several prognostic factors (55,56,57) in acute leukemia and it is clear that there are several different viewpoints and interpretations about pretreatment variables. For a good review see Cavalli (58). The therapeutic strategy is to administer combinations of drugs in order to achieve a hematologic remission. This is followed by cyclic readministration of chemotherapy to prevent leukemic relapse. Hematologic remission by induction chemotherapy is usually defined as the reduction of leukemic cells to undetectable levels and restoration of normal hemopoiesis. It is commonly agreed (59) that leukemic cell count in the bone marrow at the time of diagnosis is about 10^{12} ($\sim$1 kilogram of tumor cells).

Unfortunately, with routine methods there is no possibility to detect 10^9 or less leukemic blasts. So even in hematological remission the leukemic clone may still populate the bone marrow. Some progress has more recently been made in detecting a leukemic residue by cell culture (60,61,62) biochemical (63) and immunological (64) methods. Treatment of our patients was as follows:

AML:

induction (65):	arabinosylcytosine (Ara-C):	100 mg/m^2/day x 7
	daunorubicin :	45 mg/m^2/day x 3
maintenance:	Ara-C alternatively with : thioguanine, chlorethyl-cyclohexyl-nitrosourea, cyclophosphamide, daunorubicin.	
	(on average every 3 months)	
	mean duration of therapy 2.5 years.	

ALL:

induction:	vincristine	: 1.5 mg/m^2/weekly	x 3
	daunorubicin	: 45 mg/m^2/day	x 3
	prednisone	: 60 mg/m^2/day	x 28
	central nervous prophylaxis with metho-trexate and cranial irradiation.		
maintenance:	mercaptopurine	: 90 mg/m^2/daily	
	methotrexate	: 15 mg/m^2/weekly	
	every month reinduction therapy with vincristine and prednisone.		

From several studies of cell kinetics in acute leukemia it is clear that the proportion of leukemic cells in the proliferative cycle at any one time is low and the great majority of the leukemic cells are in the non-dividing compartment (66). The prognostic significance of the pretreatment growth characteristics of marrow and peripheral blasts is often discussed and is the cause of some controversy (67,68,69,70,71,72). Neopterin excretion is significantly increased in acute leukemias (73).
We have reported on 187 randomly selected acute adult leukemia patients with respect to the extent to which age and subtype of leukemia can be correlated with survival rate (74,75).

7.5% (14 patients) lived longer than 5 years after diagnosis and 4 of these 14 died after the 5 year survival time. Ten patients are still alive, in complete remission without therapy. Long term remissions in acute leukemia, especially in adults are seldom and early detection of relapse is difficult due to the technical problems involved in detecting a small remaining leukemic cell population. Therefore it was of particular interest to examine neopterin levels in these long term survivors and a follow up study was begun.

Results are shown in Table 12

Table 12

Neopterin levels in long term survivors with acute leukemia

patient		diagnosis	μmol neopterin / mol creatinine
Z.F.	♀	AML	224
N.M.	♀	AML	1o8
U.A.	♀	AML	6o
St.W.	♂	AML	1o5
St.I.	♀	ALL	196
T.I.	♀	ALL	112
E.M.	♀	ALL	131
L.R.	♀	PML	82
n = 8		$\bar{x}$	127
		S.D.	56
		C.V.	44
% in the pathological range			0

(Six of these patients are still alive in complete remission more than five years ($\bar{x}$ = 8.5 years) after diagnosis, one patient for 4 and one patient for 4.5 years after diagnosis).

Patients in remission but under maintenance therapy presented the following neopterin levels: 158, 2o4, 111, 199 µmol neopterin/mol creatinine. One patient with AML in relapse showed neopterin levels of 457 and of 215 µmol/mol creatinine after reinduction therapy, he died a few month later.

Polycythemia vera

Definition

Polycythemia vera is a myeloproliferative disorder, predominant in elderly patients. Recent cytogenetic data has clearly shown that polycythemia vera is a clonal disorder (76) in which a single clone is induced to proliferate thereby providing all circulating red blood cells, platelets and neutrophils. This evidence coincides with karyotypic evidence showing that chromosomal markers are shared by erythroid, myeloid and megakaryocytic progenitors. Erythropoietin levels are depressed and it seems that the disorder is not a response to a physiological stimulus but nevertheless under long term culture conditions, polycythemia vera erythroblasts respond to erythropoietin stimulation and are not autonomous (77). Therefore it appears that polycythemia vera is a clonal disorder of progenitor cells in the bone marrow, leading to increased proliferation predominantly of red cell precursors. The expanded red cell mass results in an elevated packed cell volume and increased viscosity of the whole blood. This increase in viscosity explains a higher incidence of cardiovascular and cerebrovascular disease and venous thrombosis and myeloid hyperactivity causes hyperuricemia. Extra medullary myelopoiesis contributes to enlargement of the liver and spleen. There are some criteria for making the diagnosis polycythemia vera: namely an increased red cell volume and splenomegaly, platelet count often more than 4oo,ooo/mm^3, white blood cell count more than 12,ooo/mm^3, elevated neutrophil alkaline phosphatase score or serum vitamin B_{12} level more than 9oo pg/ml (78).

Management of the disease: Phlebotomy is the most direct method, lowering the hematocrit to levels below o,5o. If, however, phlebotomy is inadequate, then either ^{32}P or alkylating agents (e.g. busulfan) are used. The mean survival time is a little more than 1o years. The development of acute leukemia in some patients is well established. Alkylating agents or - less often - ^{32}P - therapy causes this transformation (79).

Results are shown in Table 13

Table 13

Neopterin levels in patients with polycythemia vera

patient		hepato-spleno-megaly	osteomyelo-sclerosis	disease mild	severe	µmol neopterin / mol creatinine
L.M.	♀	(+)	-	+	-	319
D.V.	♀	(+)	-	+	-	278
K.O.	♂	-	-	+	-	299
K.A.	♂	-	-	+	-	284
B.A.	♀	(+)	-	+	-	186
K.S.	♀	(+)	-	+	-	234
F.F.	♂	-	-	+	-	138
Z.M.	♀	+	(+)	+	-	193
W.L.	♀	+	-	-	+	429
P.F.	♀	(+)	-	-	(+)	38o
H.M.	♀	+	+	+	-	353
H.P.	♀	++	(+)	-	+	525
L.A.	♀	++	+	-	+	81o
n = 13				neopterin $\bar{x}$		341
				S.D.		176
				C.V.		52
% in the pathological range						70

Osteomyelosclerosis,also a chronic myeloproliferative disorder is another late sequence of polycythemia vera (79). Most of our patients have received sufficient treatment so that extremely high cell counts are seldom found. The lowering of iron plasma levels by repeated phlebotomy reduces the proliferating pool of red blood cells. Neopterin levels are seldom elevated or influenced by a high proliferation capacity. It is of interest that osteomyelofibrosis give raise to higher neopterin levels(78).

Hemolytic anemias

Definition

Results are shown in Table 14

Table 14

Neopterin levels in patients with hemolytic anemias

patients: n = 16	neopterin (µmol) creatinine (mol) $\bar{x}$ = 178 S.D. = 8o C.V. = 45
autoimmune hemolytic anemia (warm autoantibody type): n = 9	$\bar{x}$ = 168
patients under immunsuppresive therapy: n = 6	$\bar{x}$ = 191
patient after splenectomy (in complete remission)	132

In general the extent of anemia is determined both by the rate of red blood cell destruction and by the capacity of the bone marrow for erythroid production. Erythrocyte defects that can cause hemolysis may be classified as either extrinsic or intrinsic hemolytic anemias. Intracorpuscular erythrocyte defects are often membrane defects, abnormalities of cell metabolism including disorders involving hemoglobin. Extracorpuscular erythrocyte defects are mostly immune and autoimmune hemolysis (80). We have chosen hemolytic anemias for neopterin measurements, because these disorders often show a hyperregenerative bone marrow which compensates the reduced lifetime of erythrocytes. This may be an example of a nonmalignant highproliferating system.

Discussion

Our investigations have shown that a high percentage of hematological neoplasias have neopterin levels in the pathological range. Most of the disorders we investigated were B-cell neoplasias (multiple myeloma, most of the non-Hodgkin lymphomas, chronic lymphocytic leukemia) but values in the pathological range were also found in patients with Morbus Hodgkin, in other neoplasias involving the lymphatic system and in myeloproliferative disorders with leukemic blood profile (chronic myelocytic leukemia, osteomyelosclerosis, polycythemia vera). Constant high neopterin levels were found in patients with advanced malignant disorders thus there is some relationship to the stage of the disease and perhaps also to tumor cell mass if it is assumed that tumor cell mass correlates with the stage of disease. There are two possible explanations for enhanced neopterin levels in hematological neoplasias: either the neopterin is produced by the malignant cell itself (perhaps as a result of dedifferentiation) or it is produced by the host organism in response to the transformed cell. Because each of these neoplasias is characterized by a different target cell of malignant

transformation, the kinetic behaviour of these cells is also different (growth fraction, duration of G_1, S_1, G_2 and mitotic phase, generation time, percentage of cells in Go). Slowly and rapidly growing tumor cells have only one common characteristic and that is the maturation defect which can occur early or late in the development of the cell (the latter is the case in most of the tumors we described). Can a maturation defect, regardless of the cell type, lead to elevated neopterin production?
This question still remains unclear. Another possibility is that neopterin is only the "product" of a pteridine precursor and that only certain cell types are able to form and/or excrete it. However, since dihydroneopterin is a necessary precursor for tetrahydrobiopterin in the cell (as a cofactor for hydroxylation reactions of phenylalanine, tyrosine, tryptophan), a certain "average value" (ground level) of precursor in all cells is assumed. Another question is whether the kidney can modulate serum neopterin levels or is capable of oxygenating dihydroneopterin to neopterin. Elevated neopterin levels in case of T-lymphocytes cannot simply be explained by proliferation because lectin stimulated T-lymphocytes showed no elevated neopterin levels. However, in vivo, stimulation of T-cells normally occurs in response to altered cell surface structures which may result from infection (through virus or microorganismus), allografts or tumor cells. It was shown (81) that this type of stimulation of T-lymphocytes both in vitro and in vivo is associated with release of neopterin into culture media or urine respectively. The possibility has been discussed, that T-cells may be genetically (stem cell concept) and/or functionally altered in B-cell neoplasias and that B-cells may be involved in the pathological process of CML. In simple terms, there are two factors influencing the response of the host to tumor cells: the antigenic "potential" of the tumor cell and the potential and capacity of the immune system. Some years ago the tumor-host relationship was regarded as passive on the part of the host and active on the part of the tumor. In recent years,

the existence of specific antigens on tumors which have been induced by chemical, physical and viral agents (mostly experiments in animals) has been thought to be the rule rather than the exception (82) and human leukemia-associated antigens detected by heteroantisera have been described since 1975 (83). The tumor cell surface is at present a focus of research attention since it is believed that alterations in the membrane may not only account for the changes in behavioural characteristics of tumor cells but also provide targets for immune recognition and destruction of these cells (84). At least three major classes of lymphoid cells are regularly involved in the immune response to tumor antigens - T-lymphocytes, B-lymphocytes and bone marrow derived cells which include natural killer cells (NK) and cells responsible for antibody-dependent cell mediated cytotoxicity (ADCC) (85). T-cells are the main cell type involved in the cell-mediated functions such as recognition and rejection of cells bearing foreign or non-self antigens. The connections between T-cell proliferation and neopterin (81) opens a new field of interpretation and virus diseases (often masked in hematological neoplasias) must be carefully excluded as delayed-type allergic reactions (contact dermatitis). Sensitivity and specificity as with other tumor markers is not 100% but is much better than is obtained with most of them. Direct estimations of neopterin in serum, or even better in the tumor cells themselves, may provide a better understanding of the pathobiochemistry of neopterin. In spite of the limitations of this assay with respect to specificty, our results have indicated that neopterin is a good "marker" for hematological neoplasias. It is at least as valuable a parameter as other clinical-chemical parameters and is particularly useful in day-to-day observation of tumor-patients and as an indicator of remission. Thus determination of altered neopterin levels is of more than mere academic interest. This is demonstrated in Table 16.

Table 16

Neopterin levels in healthy individuals, patients with hematologic neoplasias, patients with hematologic neoplasias in remission and patients with benign hematologic disorders.

	n	µmol neopterin / mol creatinine		% in the pathological range
healthy individulas	417	$\bar{x}$	128	o
		S.D.	4o	
		C.V.	31	
benign hematological disorders [1]	25	$\bar{x}$	174	16
		S.D.	72	
		C.V.	41	
hematologic neoplasias [2]	141	$\bar{x}$	52o	85
		S.D.	33o	
		C.V.	63	
hematological neoplasias [3] in remission	52	$\bar{x}$	193	12.5
		S.D.	85	
		C.V.	44	

1) autoimmunohemolytic anemia n = 16, benign paraproteinaemia n = 9.

2) Morbus Hodgkin n = 16, non-Hodgkin lymphomas n = 33, chronic myelogenous leukemia n = 21, chronic lymphocytic leukemia n = 26, multiple myeloma n = 27, polycythemia vera n = 13, acute leukemia n = 5.

3) Morbus Hodgkin n = 24, non-Hodgkin lymphomas n = 13, multiple myeloma n = 7, acute leukemia n = 8.

References

1. Kaufman, S.: Ann.Rev.Biochem. 36, 117-184 (1967).

2. Wachter, H., Hausen, A., Reider, E., Schweiger, M.: Naturwissenschaften 67, 61o-611 (198o).

3. Wachter, H., Graßmayr, K., Hausen, A.: Cancer Lett. 6, 61-66 (1979).

4. Wachter, H., Hausen, A., Graßmayr, K.: Hoppe-Seyler's Z. Physiol. Chem. 36o, 1957-196o (1979).

5. Rokos, H., Rokos, K., Frisius, H., Kirstaedter, J.H.: Clin. Chem. Acta 1o5, 275-286 (198o).

6. Fuchs, D., Hausen, A., Huber, Ch., Margreiter, R., Reibnegger, G., Spielberger, M., Wachter, H.: Hoppe-Seyler's Z. Physiol. Chem., 363, 661-664 (1982).

7. Lieberman, G.J.: Industrial Quality Control, 14, 7-9 (1958).

8. Huber, Ch., Michlmayr, G., Falkensamer, M., Fink, U., zur Nedden, G., Braunsteiner, H., Huber, H.: Clin. exp. Immunol. 21, 47-53 (1975).

9. Michlmayr, G.: Fortschritte der Medizin 96 (38), 1928-1932 (1978).

1o. Shiftan, T.A., Mendelsohn, J.: Blut 38, 1-8 (1979).

11. De Vita, V.T., Serpik, A.A., Carbone, P.P.: Ann. Intern. Med. 73, 881-895 (197o).

12. Santoro, A., Bonadonna, G.: Cancer Chemother. Pharmacol. 2, 1o1-1o5 (1979).

13. Valagussa, P., Santoro, A., Kenda, R., Fossati Bellani, F., Franchi, F., Banfi, A., Rilke, F., Bonadonna, G.: Br. Med. J. 28o, 216-219 (198o).

14. Rosner, F., Grünwald, H.W.: Am. J. Med. 58, 339-353 (1975).

15. Carbone, P.P., Kaplan, H.S., Musshoff, K., Smithers, D.W., Tubiana, M.: Cancer Res. 31, 186o-1861 (1971).

16. Lennert, K., Mohri, N., Stein, H., Kaiserling, E., Müller-Hermelink, H.K.: Malignant lymphomas, Springer Verlag, Berlin-Heidelberg-New York (1978).

17. Gerard-Marchant, R., Hamlin, I., Lennert, K., Rilke, F., Stansfeld, A.G., van Unnik, J.A.M.: Lancet II, 4o6-4o8 (1974).

18. Huber, H., Gattringer, C.: Arch. Geschwulstforsch. 49 (8), 694-7o1 (1979).

19. Ferraris, A.M., Giunti, P., Gaetani, G.F.: Blood 54, 928-932 (1979).

2o. Rosenberg, S.A.: N. Engl. J. Med. 3o1, 924-928 (1979).

21. Huber, Ch., Michlmayr, G., Huber, H., Braunsteiner, H.: Dtsch. Med. Wschr. 49, 1795-1799 (1977).

22. Stacher, A., Heinz, R., Waldner, R., Böhnel, J.: Wien. klin. Wschr. 92, 52o-525 (198o).

23. Silvestrini, R., Piazza, R., Riccardi, A., Rilke, F.: J.Natl. Cancer Inst. 58, 499-5o4 (1977).

24. Bremer, K.: Recent results in Cancer Research, 65, 5-11 (1978).

25. Sauter, Chr., Sulser, H.: Die malignen Lymphome, In: Internistische Krebstherapie. Brunner, K.W., Nagl, G.A. (eds) 2. Auflage, 25o-276 (1979).

26. Fialkow, P.J., Jacobson, R.J., Papayannopoulou, T.: Am. J. Med. 63, 125-129 (1977).

27. Fialkow, P.J., Jacobson, R.J., Singer, J.W., Sacher, R.A., McGuffin, R.W., Neefe, J.R.: Blood 56, 7o-73 (198o)

28. Martin, J.P., Najfeld, V., Hansen, J.A., Penfold, G.K., Jacobson, R.J., Fialkow, P.J.: Nature 287, 49-5o (198o).

29. Maurice, P.: Die chronischen Leukämien des Erwachsenen. In: Internistische Krebstherapie. Brunner, K.W., Nagel, G.A. (eds) Springer Verlag, Berlin-Heidelberg-New York, p.: 229-249 (1979).

3o. Gavosto, F.: Cell Tissue Kinet. 7, 151-163 (1974).

31. DBM Study Group: Eur. J. Cancer 9, 583-589 (1973).

32. Haanen, C., Hellriegel, K.P., Machins, D.: Europ. J. Cancer 15, 8o3-8o9 (1979).

33. Schilling, R.F., Crowley, J.J.: Am. J. Hematol. 7, 1-1o (1979).

34. Collins, R.K., Jackson, F.F., Morrison, F.S., Meydrech, E.F.: Southern Med. J. 76, 642-645 (1979).

35. Schöch, G., Garbrecht, M., Heller-Schöch, G., Baisch, H., Leifer, W.: Blut 38, 391-396 (1979).

36. Dameshek, W.: Blood 29, 566-584 (Suppl.) (1967).

37. Fialkow, P., Najfeld, V., Reddy, A., Singer, J., Steinmann, L.: Lancet 2, 444-446 (1978).

38. Huber, H., Michlmayr, G., Huber, Ch., Falkensamer, M.: Klin.Wschr. 54, 699-7o8 (1976).

39. Theml, H., Love, R., Begemann, H.: Ann. Rev. Med. 28, 131-141 (1977).

4o. Baccarani, M., Killman, S.A.: Scand. J. Haematol. 9, 283-292 (1972).

41. Stryckmans, P.A., Debusscher, L., Sacquet, M.: Cancer Res. 36, 3o34-3o38 (1976).

42. Stryckmans, P.A., Debusscher, L., Collard, E.: Clin. Haematol. 6, 159-165 (1977).

43. Moayeri, H., Sokal, J.E.: Am. J. Med. 66, 773-778 (1979).

44. Van der Woerd de Lange, J.A., Döhrmann, J., Huber, Ch., Schick, P., Ravert, K., Begemann, H.: Blut 37, 319-326 (1978).

45. Rai, K.R., Sawitsky, A., Cronkite, E.P., Chanana, A.D., Levy, R.N., Pasternack, B.A.: Blood 46, 219-234 (1975).

46. Knospe, W.H., Loeb, V., Huguley, C.M.: Cancer 33, 555-562 (1974).

47. Barton, R.W., Goldschneider, I.: Molecular & Cellular Biochemistry 28, 135-147 (1979).

48. Durie, B.G.M., Salmon, S.E.: Cancer 36, 842-854 (1975).

49. Salmon, S.E., Smith, B.A.: J. Clin. Invest. 49, 1114-1121 (197o).

5o. Durie, B.G.M., Salmon, S.E., Moon, T.E.: Blood 55, 364-372 (198o).

51. Mill, W.B., Griffith, R.: Cancer 45, 647-652 (198o).

52. Alexanian, R., Salmon, S.E., Bonnet, J., Gehan, E., Haut, A., Weick, J.: Cancer 4o, 2765-2771 (1977).

53. McIntyre, O.R.: N. Engl. J. Med. 3o1, 193-196 (1979).

54. Clarkson, B.D.: Human tumor cell kinetics, Monograph No. 3o, 81-12o, Bethesda, National Cancer Institute (1969).

55. Brandmann, H., Bukowski, R.M., Greenstreet, R., Hewlett, J.S., Hoffman, G.C.: Cancer 44, 1o62-1o65 (1979).

56. Bernard, J., Jacquillat, C., Weil, M., Chastang, C., Gémon-Auclerc, M.F.: Actual Hématol. 9me série, 3-12 (1975).

57. Keating, M.J., Smith, T.L., Gehan, E.A., McCredie, K.B., Bodey, G.P., Spitzer, G., Hersh, E., Gutterman, J., Freireich, E.J.: Cancer 45, 2o17-2o29 (198o).

58. Cavalli, F.: Prognostische Faktoren und Therapie der akuten Leukämien beim Erwachsenen, Verlag Hans Huber, Bern-Stuttgart-Wien (198o).

59. Killman, S.: Sem. Hematol. 1, 38-1o2 (1968).

6o. Broxmeyer, H.E., Grossbard, E., Jacobson, N., Moore, M.A.S.: New. Engl. J. Med. 3o1, 346-351 (1979).

61. Broxmeyer, H.E., Ralph, P., Margolis, V.B., Nakainz, I., Meyers, P., Kapoor, N., Moore, M.A.S.: Leukemia Research 3, 193-2o3 (1979).

62. Broxmeyer, H.E., Jakobsen, N., Kurland, J., Mendelsohn, N., Moore, M.A.S.: I. Nat. Cancer Inst. 6o, 497-511 (1978).

63. Viola, M.V., Frazier, M.: New Engl. J. Med. 294, 75-8o (1976).

64. Baker, M.A., Falk, J.A., Carter, W.H., Taub, R.N. and the Toronto Leukemia Study Group: New Engl. J. Med. 3o1, 1353-1357 (1979).

65. Rai, K.R., Holland, J.F., Glidwell, O.J.: Proc. Am. Assoc. Cancer Res. 16, 265-272 (1975).

66. Killman, S.A.: Clinics in Haematology 1, 95-113 (1972).

67. Amadori, S., Petti, M.C., Mastrovincenzo, C., Testa, M.G.: Acta hemat. 61, 237-242 (1979).

68. MacKinney, A.A., Flynn, B.: Am.J. Haematol. 5, 93-99 (1978).

69. Kremer, W.B., Vogler, W.R., Chan, Y.K.: Cancer 37, 39o-4o3 (1976).

7o. Amadori, S., Petti, M.C., Francesco, A., Chierichini, A., Mastrovincenzo, C., Testa, M.G., Mandelli, F.: Cancer 41, 1154-116o (1978).

71. Murphy, S.B., Aur, R.J.A., Simone, J.V., George, S., Maur, A.M.: Blood 49, 683-691 (1977).

72. Hart, J.S., George, S.L., Frei III, E., Bodey, G.P., Nickerson, R.C., Freireich, E.J.: Cancer 39, 16o3-1617 (1977).

73. Kostron, C., Fuchs, D., Hausen, A., Reibnegger, G., Wachter, H.: In: Biochemical and Clinical Aspects of Pteridines. First Workshop on Pteridines. St.Christoph, Arlberg (Austria), March 7-1o, 1982, pp.157-171, Eds.: H.Wachter, H.Ch.Curtius, W.Pfleiderer. Walter de Gruyter, Berlin-New York (1982).

74. Grünewald, K., Abbrederis, K., Mittermaier, P., Huber, H.: Blut 44, 289-296 (1982).

75. Grünewald, K., Abbrederis, K., Huber, H.: Onkologie 4, 28o (1981).

76. Adamson, J.W., Fialkow, P.J., Murphy, S., Prchal, J.F., Steinmann, L.: New Engl. J. Med. 295, 913-916 (1976).

77. Zanjani, E.D.: Semin. Hematol. 13, 1-33 (1976).

78. Wintrobe, M.M.: Polycythemia vera, In: Clinical Haematology. 7th edn., 988-1oo8, Philadelphia: Lea and Febinger (1975).

79. Landaw, S.A.: Semin. Haematol. 13, 33 (1976).

8o. Begemann, H., Harwerth, H.G.: Hämolytische Anämien, In: Praktische Hämatologie. 7th edn., 11o-157, Georg Thieme Verlag Stuttgart (1977).

81. Huber, Ch., Margreiter, R., Spielberger, M., Fuchs, D., Hausen, A., Reibnegger, G., Wachter, H.: In: Biochemical and Clinical Aspects of Pteridines. First Workshop on Pteridines. St.Christoph, Arlberg (Austria), March 7-1o, 1982, pp. 227-232, Eds.: H.Wachter, H.Ch.Curtius, W. Pfleiderer. Walter de Gruyter, Berlin-New York (1982).

82. Alexander, P.: Tumor immunology. In: The immunsystem. Hobart, M.J., McConnel, I. (eds). Blackwell scientific publications. Oxford-London-Edinburgh-Melbourne. p.: 296-307 (1975).

83. Greaves, M.F., Brown, G., Rapson, N.T., Lister, T.A.: Clin. Immunol. Immunopathol.: 4, 67-84 (1975).

84. Hoog, N.: The immune response to C-type virus induced tumors. In: Surfaces of normal and malignant cells. Hynes, R.O. (ed). John Wiley & Sons. Chichester-New York-Brisbane-Toronto. p.: 287-322 (1979).

85. Weissman, I., Hood, L., Wood, W.: Essential concepts in immunology. In: Benjamin/Cummings, Mento Park (1978).

URINARY NEOPTERIN IN PATIENTS WITH HEAD AND NECK CANCER

G. Reibnegger, D. Fuchs, A. Hausen, H. Wachter
Institute of Medical Chemistry and Biochemistry
University of Innsbruck,
Innsbruck, Austria

E. Bichler and K. Böheim
ENT-Department, University of Innsbruck,
Innsbruck, Austria

Introduction

It has now been known for several years that the concentration of neopterin in urine of patients with various malignancies and with viral infections is significantly increased as compared to the urine of healthy persons (1-4).

Recently results have been reported on investigations directed towards more homogeneous groups of malignant tumors by our laboratory: Studies on women with cancer of the genitals (5) as well as in patients with lymphoid and myeloid neoplasms (6) both indicated that urinary neopterin sampling may be of great value as an additional parameter in diagnosing those tumors and especially in monitoring therapy by long term studies.

However, the investigation concerning gynaecological tumors revealed an interesting and unexpected feature: While in patients with ovarian adenocarcinomas, which are mostly composed of undifferentiated cells, correlation between positive clinical tumor diagnosis and elevation of urinary neopterin levels was quite satisfying, this was not the case in women

Biochemical and Clinical Aspects of Pteridines · Volume 1
Cancer · Immunology · Metabolic Diseases

with squamous cell carcinomas of the cervix uteri, predominantly highly differentiated tumors, where frequency of elevated neopterin in urine was considerably smaller.

The present paper is designed to report preliminary results of a further study on a homogeneous group of malignancies: We measured urinary neopterin in patients with various types of head and neck cancer in order to get some more knowledge of the behaviour of this new parameter in assessment of neoplastic diseases. Our second intention was to examine a group of malignant tumors, which in most cases are representative examples of squamous cell carcinomas, to further clarify the above mentioned observation of different properties of this type of cancer with respect to pteridine excretion in urine of patients.

Patients and Method

Urinary neopterin was determined in urine of 35 patients with head and neck cancer. 32 of the patients were male, 3 female. Their ages ranged from 4o to 78 years, median age being 58 years.

The localisation of the tumors was as follows: 7 in the larynx, 8 in the oral cavity, 13 in the pharynx and 4 in the paranasal sinuses. 3 of the patients showed isolated cervical metastases.

Pathohistologically 33 of the neoplasms were classified as squamous cell carcinomas, mostly highly differentiated, one tumor was a lymphoepithelial carcinoma ("Schmincke-Regaud") and one a rhabdomyosarcoma.

In 3o of the patients either the diagnosis had been made immediately before neopterin sampling and hence the treatment of the tumor had not begun at the time of the investigation or medication had been interrupted for at least more than 3 months previously.

Further samples of 11 of these and 4 other patients were analysed 12 days after the first out of three polychemotherapeutic treatments with a combination of cis-dichlorodiamine-platinum-(II), bleomycin and amethopterin.

The patient with the rhabdomyosarcoma received another kind of chemotherapy-cyclophosphamide, adriamycin, vincristine and 5-(3,3-dimethyl-1-triazenyl)-1H-imidazole-4-carboxamide (DTIC).

In all cases, urine samples collected in the morning were used for analysis. The samples were analysed either immediately or frozen at once and stored at -2o^{o}C in the dark to prevent oxidative or photochemically induced alterations in the composition. Analyses were performed without knowledge of the clinical diagnosis.

In order to correct for physiological variations of the urinary concentration, neopterin values were related to simultaneously determined creatinine levels.

As in a previous study (8) neopterin values of more than 25o umol per mol creatinine were considered to be elevated; this corresponds to the mean plus three standard deviations in a group of 79 apparently healthy persons. Due to the preliminary character of this report division of the patients into sex and age groups was not performed.

For analysis the high performance liquid chromatography (HPLC)-technique was employed. The procedure of determining neopterin and creatinine simultaneously in urine samples has been described previously (7,9).

Results

The levels of urinary neopterin, related to creatinine, found in the group of 3o untreated patients with various types of head and neck cancer are demonstrated in figure 1. The status of differentiation of the respective tumor cells is also indicated in the figure.

As can be seen, the frequency of elevated neopterin levels is surprisingly low: Only 7 of the patients show increased values in contrast to 23 individuals with neopterin concentrations within the normal range.

The same results can be observed in the group of 15 individuals 12 days after the start of a set of three chemotherapeutic cures - shown in figure 2: Only four patients exhibit raised urinary neopterin levels.

In the patient with the rhabdomyosarcoma, who is not represented in the figures, the value was 171 µmol neopterin per mol creatinine.

Both in figures 1 and 2 no clear correlation can be observed between differentiation status of the tumor tissue and neopterin excretion in urine.

Discussion

Obviously in head and neck cancer urinary neopterin levels are - in general - not significantly raised in contrast to healthy subjects, as is the case with other types of cancer. Thus, the above mentioned divergent behaviour of gynaecological squamous cell carcinomas in comparison to

Figure 1 : Neopterin levels in first morning urine samples of patients with malignant tumors of the head and neck region prior to the onset of therapy

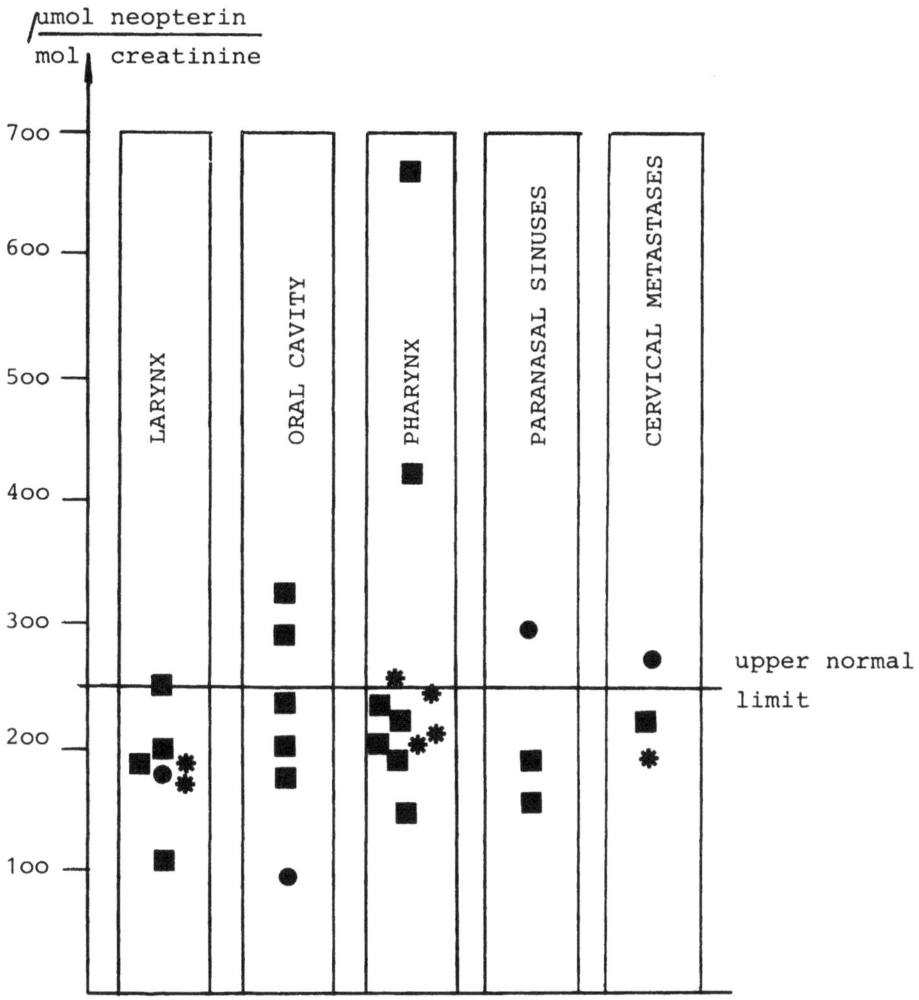

■ highly differentiated tumor
✱ undifferentiated tumor
● unknown state of differentiation

Figure 2 : Neopterin levels in first morning urine of patients with head and neck cancer 12 days after the first out of three chemotherapeutic cures

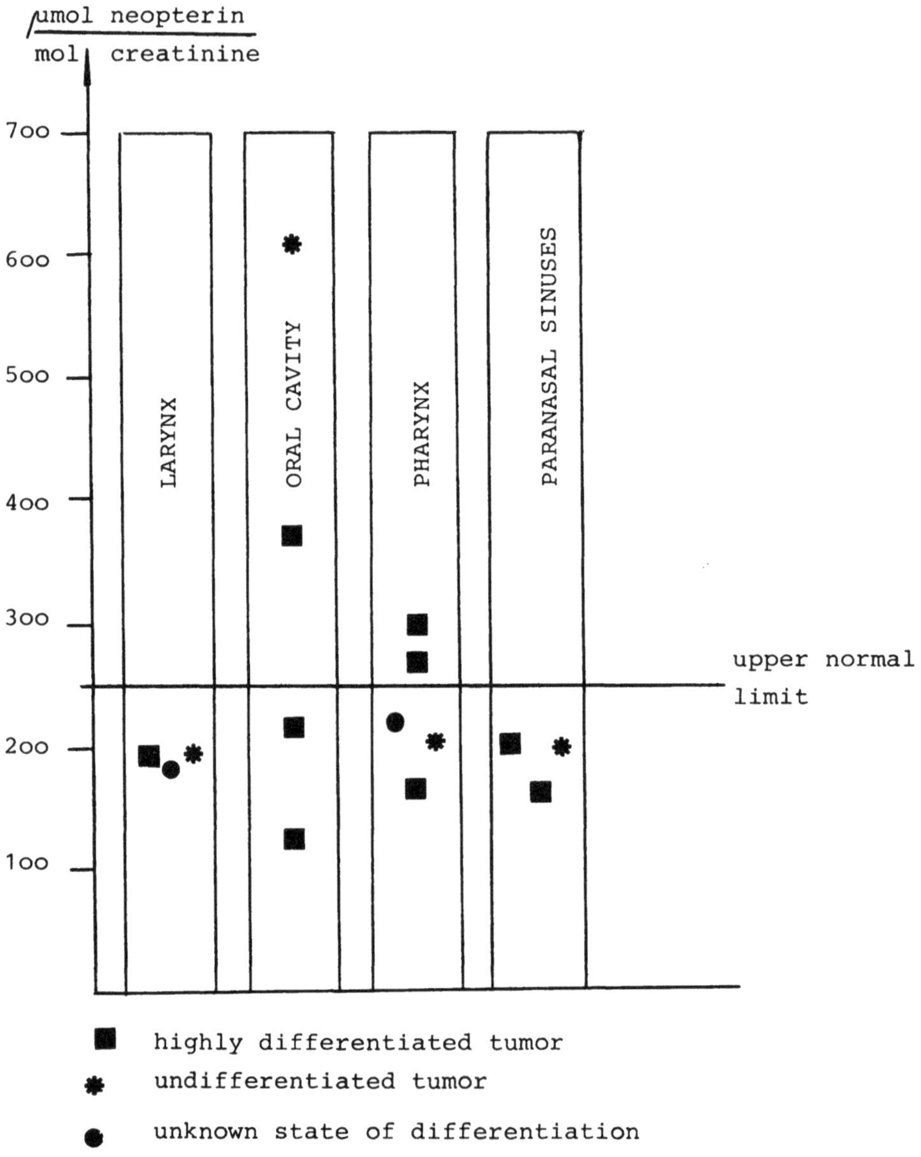

ovarian adenocarcinomas and haematological neoplasms is also detectable in tumors of the same histological class localised elsewhere: Squamous cell carcinomas do not evoke such elevated neopterin excretion as haematological neoplasms or ovarian carcinomas do.
We feel, that this statement is justified despite of the relatively small group of patients investigated hitherto. Our measured values do not seem to be an artefact due to an accidental overrepresentation of malignancies in a premature state: most of our patients had rather advanced tumors, classifiable as T3NØ,1MØ according to the TNM-system.

An interesting feature can be shown by detailed analysis of the results shown in the two figures: 11 of the 15 patients in figure 2 - 12 days after chemotherapy - are also represented in figure 1 - according to absence of medication. In 9 of them the neopterin levels are higher after therapy in contrast to the values before treatment. The number of patients is too small to decide definitely if this result is statistically significant, but in connection with other results obtained in patients with graft rejection reactions (1o) and with follow up studies in gynaecological cancer (6), we believe this to be true. Due to the observations cited above we believe that this behaviour might be caused by recovery of the lymphocytes from the treatment.

We could not find any clear correlation between neopterin levels in urine of the patients on the one hand and differentiation state of the tumor tissues on the other hand, as can be seen from the figures. However - some sort of relationship is suggested by the fact of a generally different behaviour of squamous cell carcinomas. An exact decision on this important problem is not possible at the present moment: Our patient group has been too small up to now to allow a statistically corroborated answer, and there is the difficulty of assigning a certain degree of differentiation to a tumor in a global sense, which prohibits general validity of the results.

Summary

In contrast to other malignancies head and neck cancer seems not to cause similar elevations of urinary neopterin. Whether a relationship between differentiation status of the neoplasm and neopterin excretions in urine of the patients exists or not, is not decidable at the moment. Further investigations concerning this question seem to be of considerable interest, especially in order to get deeper insight into the exact nature of neopterin excretion in urine.

References

1. Wachter, H., Hausen, A., Grassmayr, K.: Ber.ÖGKC 2, 9o-91 (1979).

2. Wachter, H., Hausen, A., Grassmayr, K.: Hoppe Seyler's Z.Physiol.Chem. 36o, 1957-196o (1979).

3. Rokos, H., Rokos, K., Frisius, H., Kirstaedter, H.J.: Clin.Chim.Acta 1o5, 275-286 (198o).

4. Stea, B., Halpern, R.M., Halpern, B.C., Smith, R.A.: Clin.Chim.Acta 113, 231-242 (1981).

5. Hausen, A., Fuchs, D., Grünewald, K., Huber, H., König, K., Wachter, H.: Clin.Biochem. 15, 34-37 (1982).

6. Bichler, A., Fuchs, D., Hausen, A., Hetzel, H., König, K., Wachter, H.: Clin.Biochem. 15, 38-4o (1982).

7. Hausen, A., Fuchs, D., König, K., Wachter, H.: J.Chromatogr. 227, 61-7o (1982).

8. Hausen, A., Fuchs, D., Grünewald, K., Huber, H., König, K., Wachter, H.: Clin.Chim.Acta 117, 297-3o5 (1981).

9. Hausen, A., Fuchs, D., Wachter, H.: J.Clin.Chem.Clin. Biochem. 16, 375-378 (1981).

1o. Fuchs, D., Hausen, A., Huber, C., Margreiter, R., Reibnegger, G., Spielberger, M., Wachter, H.: Hoppe Seyler's Z.Physiol.Chem. 363, 661-664 (1982).

A PTERIDINE BINDING VARIANT OF ALPHA$_1$-ACID GLYCOPROTEIN, ACCUMULATED IN THE BLOOD OF PATIENTS WITH MALIGNANT DISEASES

Irmgard Ziegler

Institut für Toxikologie und Biochemie der Gesellschaft für Strahlen-und Umweltforschung, Abteilung Zellchemie; D-8 München 2, Germany

Konrad Maier

Institut für Toxikologie und Biochemie der Gesellschaft für Strahlen-und Umweltforschung, Abteilung Enzymchemie; D-8042 Neuherberg, Germany

Wulf Armarego

Medical Research Group, The John Curtin School of Medical Research, The Australian National University; Canberra City, Australia

Michael Fink

III.Medizinische Klinik der Universität, Klinikum Großhadern; D-8 München 70, Germany

Introduction

During leukemic proliferation highly increased biopterin concentrations were found to be characteristic for blasts (1). Increased plasma levels only resulted from impaired excretion (1,2). The question remains whether in cases of solid tumors cell proliferation is indicated by increased levels of plasma biopterin.

Increases in protein bound tyrosine in the supernatant fraction from blood extract with digitonin followed by alcohol and TCA precipitation had been considered as indicative of neoplastic diseases (3). Its identification and relation

Biochemical and Clinical Aspects of Pteridines · Volume 1
Cancer · Immunology · Metabolic Diseases

to pteridine metabolism is outlined below. It has been described in detail in earlier publications (4,5,6).

Materials and Methods

Details of blood sampling, isolation and analysis (polyacrylamide gel electrophoresis, electrofocusing, amino acid analysis, Crithidia assay and quantitative determination) are described in detail in (4,5). The experimental conditions for the binding studies and the sialic acid analysis are given in (6).

Results

1) Biopterin levels in the blood of patients with solid tumors. Normally, in 1 ml blood 3.0 (S.D.$\pm$ 0.66) ng biopterin are found. In contrast to leukemias (1), in a total of 132 blood samples, drawn from patients with lymphomas or solid tumors (including 9 cases with repeated drawings) in only 22 could increased biopterin levels (3.9 - 14.0 ng /ml blood) be found. In most of them a concomitant, even slight increase in creatinine indicated an impaired excretion.

   In an earlier investigation (2), a close correlation between creatinine and biopterin retention during renal insufficiency had been obtained. Longitudinal sections revealed that in cases of solid tumors, the biopterin attained higher levels than would be expected from impaired excretion alone (fig.1). In those cases possibly, an overproduction was manifested.

   In view of these complex correlations, plasma biopterin does not appear to be a reliable and clear cut indicator for lymphomas and solid tumors.

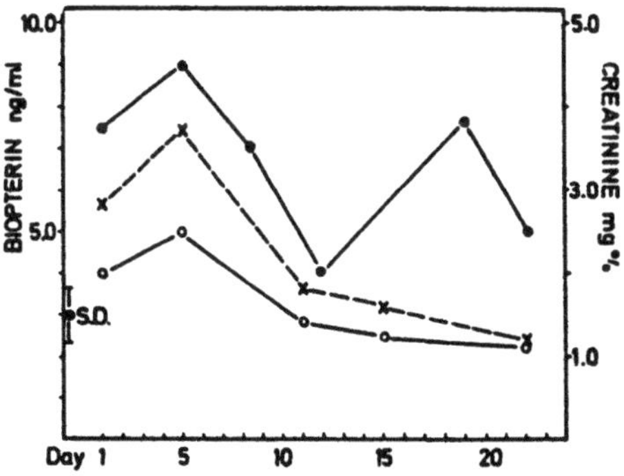

Fig.1 Biopterin and creatinine levels of blood during a case of rhabdomyosarcoma. Initial period with ureteral compression. The second peak coincided with a period of febril pyelonephritis. mean and ± S.D. of biopterin. ●—● actual biopterin; x---x creatinine; o—o biopterin as to be calculated from the correlation curve with creatinine (see ref.2)

2) Isolation and characterization of a pteridine binding variant of alpha$_1$-acid glycoprotein *. The protein under investigation was isolated by Concanavalin A affinity chromatography. The molecular weight as obtained by means of SDS polyacrylamide gel electrophoresis, the amino acid composition, and the amounts of sugars were identical with a commercial preparation of AGP*. A number of data however, indicated that the glycoprotein in our hands did not represent the total of AGP but rather a variant, AGP_C* and AGP_M*, respectively. The main differences can be outlined as the following ones.

Polyacrylamide gel electrophoresis without SDS and electrofocusing revealed the presence of microheterogeneic bands in AGP_M, less negatively charged than AGP_C or AGP (4).

The higher p_I values of AGP_M are not due to a decreased

- - - - - - - - - - - - -

Abbreviations:

AGP = alpha$_1$-acid glycoprotein, total of the microheterogeneic variants. In the experiments a preparation from Sigma (St.Louis) was used, which was homogenous in SDS polyacrylamide gel electrophoresis

AGP_M = variant of AGP, isolated from blood of patients with malignant diseases

AGP_C = variant of AGP, isolated from blood of controls

percentage of sialic acid. They rather may be explained by the protonation and binding of the chromophoric group which can be split from AGP_M by reductive treatment, by guanidine·HCl (4) or by neuraminidase treatment (6).

E_{280} (1%; 1 cm) of AGP was found to be 9.0 (see also ref.7), whereas AGP_M averages 16.0. The O.R.D.spectra at 260 - 310 nm reveal marked increases in chirality in AGP_M and AGP_C, as compared with AGP. The chirality found for AGP_M in the 320-400 nm region lacks AGP_C and AGP. This indicates different conformations of the glycoprotein.

AGP_M is characterized by a chromophoric group (fig.2). Its spectral data and its low growth factor activity towards Crithidia fasciculata indicated that it is a pteridine without the specific side chain at C(6).

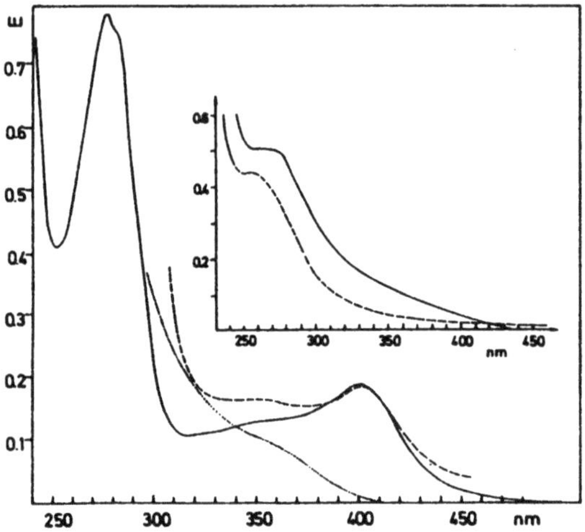

Fig.2 Absorption spectrum of Con A-reactive AGP_M and AGP_C and the effect of reductive treatment. ——— Con A-reactive AGP_M (380 ug/ml); 100 mM ammoniumacetate (pH 6.7); - - - - 1 min after addition of sodium dithionite; ···· 4 min after addition of sodium dithionite. Inset: ——— Con A-reactive AGP_C in 100 mM ammoniumacetate (pH 6.7; qualitatively); -- - - - AGP_M after reduction with dithionite and subsequent separation by molecular sieving in the presence of dithionite (qualitatively). E = optical density

3) Binding studies by means of model reactions. Binding studies were performed by incubation of AGP with ^{14}C-pterins and subsequent separation of the glycoprotein by molecular sieving. They are described in detail in (6). Isoxanthopterin-^{14}C but not xanthopterin-^{14}C binds to AGP. AGP_M adds further isoxanthopterin-^{14}C.

The binding of isoxanthopterin to AGP and AGP_M is lost by previous neuraminidase treatment. This corroborates the essential role of sialic acid for the binding of the chromophoric group.

Isoxanthopterin binding results in an increase of E_{280} from 9.0 to 17.1-26.5 and an induced chirality in the 340-400 nm range. The calculated contribution of isoxanthopterin to E_{280} of the adduct is minimal; therefore the increase points to an induced change in conformation.

4) Levels of AGP_M during clinical cases. The concentration of total AGP is 700-1000 ug/ml blood (8). The variant AGP_C comprises about 10% of total. The ranges of AGP_M during various stages of malignant diseases are given in table 1.

Table 1 Concentrations of AGP, AGP_C and AGP_M in blood (ug/ml)

AGP	AGP_C	AGP_M solid tumors	AGP_M deteriorating cases of solid tumors, leukemias
700	75-105	150-500	800-1000

A screening of 800 malignant and non-malignant cases is described in detail (5). About 75% of the patients with lymphomas or solid tumors showed increased levels of AGP_M, whereas during leukemias, about 95% were positive (fig.3). In all longitudinal sections, partially followed

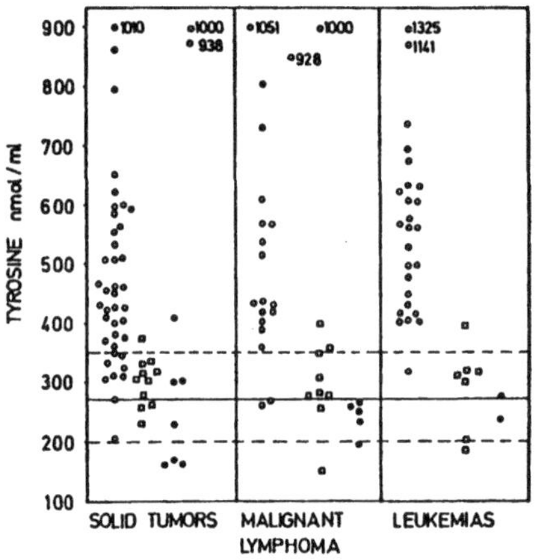

Fig.3 Levels of AGP_M during malignancies, indirectly determined by protein-bound tyrosine in Fraction b.

---- mean and ± S.D. for control subjects; O untreated and resistant to therapy; □ partial remission; ● complete remission

for 10 months, the AGP_M levels were closely correlated with the success of (mostly cytostatic) treatment.

AGP is known to be a member of acute phase reactants (8). The data of the screening classified AGP_M to some extent as a member of those, but its increase during heavy inflammations or in cases of main surgery were less drastic or more transient than those known for AGP. Thus, the clinical aspects corroborate the view that AGP_M may be related to the "abnormal orosomucoid" of Rudman (9,10), the chromophoric group of which had not been recognized before.

5) The correlation of biopterin and AGP_M levels during leukemias. During 11 longitudinal sections of leukemias, the biopterin accumulation in the leukemic cells was compared with the levels of AGP_M. As demonstrated in fig.4, in all cases changes in biopterin were clearly followed by the amounts of AGP_M. They possibly suggest

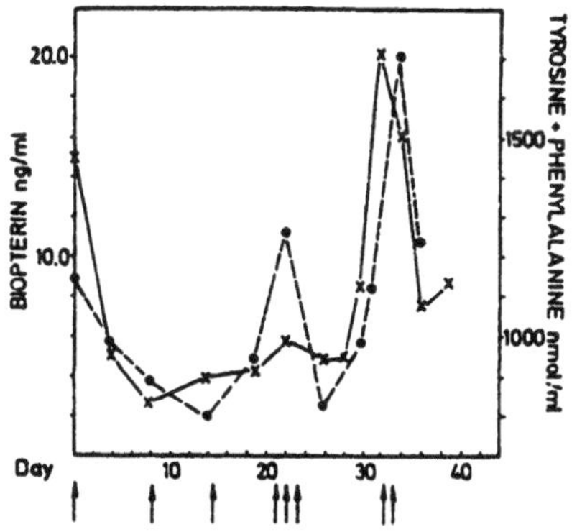

Fig.4 Levels of blood biopterin and of AGP_M during a blast crisis of chronic myelocytic leukemia. AGP_M was indirectly determined by amino acid analysis of Fraction b. x——x biopterin (mean and ± S.D. of controls: 3.0 (0.66)ng/ml blood; ●--● AGP_M (mean and ± S.D. of controls: 189.7 (49.9) nmol phenylal. + tyrosine/ml blood. ↑ cytostatic treatmant

that biopterin accumulation and increased AGP_M are closely related to each other.

A chromophoric group of a urinary microglycoprotein during a case of relapsed plasma cell leukemia has already been observed but still remained uncharacterized (11). Those glycoproteins may serve as receptors of pteridine intermediates. By their binding to AGP_M (mol. w.= 42 000 daltons), they are possibly retained instead of the pterins, which are easily excreted during normal renal function (2). Increased urinary levels of biopterin and neopterin, frequently but not consistently found during malignant growth (12,13,14) may represent the last overspill. A rapid a specific assay for AGP_M may thus develop into a tool for discriminating acute phases and malignant growth.

References

1. Ziegler,I., Fink,M., Wilmanns,W.: Blut 44, 231-240 (1982)
2. Ziegler,I., Fink,M.: Biochem.Med., in press (1982
3. Ziegler,I., Kokolis,N.: in "Chemistry and Biology of

Pteridines, p.165-170. Elsevier, North Holland; New York, Amsterdam, Oxford (1979)

4. Ziegler,I., Maier,K., Fink,M.: Cancer Research 42, 1567-1573 (1982)

5. Fink,M., Ziegler,I., Maier,K., Wilmanns,W.:Cancer Research 42, 1574-1578 (1982)

6. Ziegler,I., Armarego,W., Fink,M.: 7th Int.Symp.Pteridines and Folic acid derivatives. Chemical, Biological and Clinical Aspects. Walter de Gruyter, Berlin,New York (1982) (accepted)

7. Schmid,K.: in: The Plasma Proteins,vol.1,p.183-228, Academic Press; New York (1975)

8. Koj,A.: in: Structure and Function of Plasma Proteins, p. 73-125, Plenum Publ.Co, London (1974)

9. Rudman,D., Chawla,R.K ., Del Rio,A.E.,Hollins,B.M., Hall,E.C.,Conn,J.M.: J.Clin.Invest.54, 147-155 (1974)

10. Rudman,D., Treadwell,P.E.,Vogler,W.R., Howard,C.H., Hollins,B.: Cancer Research 32,1951-1959 (1972)

11. Seon,B.K., Pressman,D.: Biochemistry 17, 2815-2821(1978)

12. Halpern,R.M., Halpern,B.C., Stea,B., Dunlap,A.,Conklin, K., Clark,B., Ashe,H., Sperling,L., Halpern,J.A.,Hardy, D., Smith,R.A.: Proc.Natl.Acad.Sci.(USA) 74, 587-591 (1977)

13. Rokos,H., Rokos,K., Frisius,H., Kirstaedter,H.J.:Clin. Chim.Acta 105, 275-286 (1980)

14. Wachter,H., Hausen,A., Grassmayr,K.: Hoppe Seyler's Zeitschr.Physiol.Chem. 360, 1957-1960 (1979)

4

PTERIDINES IN IMMUNOLOGY

PTERIDINES IN THE ASSESSMENT OF LYMPHOCYTE ACTIVATION

Christoph Huber
Department of Internal Medicine, University of Innsbruck

Raimund Margreiter, Martin Spielberger
Department of Surgery I, University of Innsbruck

Dietmar Fuchs, Arno Hausen, Gilbert Reibnegger, Helmut Wachter
Department of Medical Chemistry, University of Innsbruck

Introduction

Recent evidence presented during this meeting suggests that cellular proliferation is invariably associated with release of distinct Pteridines (PTs). Lymphocytes during functional activation undergo clonal expansion of cells bearing specific surface receptors for the stimulating antigens. We have tested the hypothesis that during clonal expansion of allo-antigen activated lymphocytes PTs are released which are specific for this cell type.

Materials

PT levels in culture supernatants were assessed by means of high pressure liquid chromatography (HPLC) as previously described (1). Results of the evaluation of PT specificity with 4.2 min retention are expressed in arbitrary fluorescence units; those of the PT specificity with 6.7 min retention, which previously had been identified as neopterin (2), are given in ng/ml or in µmol/ml.

Biochemical and Clinical Aspects of Pteridines · Volume 1
Cancer · Immunology · Metabolic Diseases

Mixed lymphocyte cultures were established by co-cultivation of responder T cells from individual A with mitomycin-C treated stimulator cells from individual B as previously described (3). Cultures of A cells with autologous and mitomycin-C treated stimulator cells and cultures of A cells alone served as controls. At times indicated cells were assessed for proliferation by ^{3}H-thymidine incorporation (3) and supernatants for PT levels.

Results

The first set of experiments demonstrated that in vitro stimulation of T lymphocytes with allogeneic but not with autologous cells leads to a specific production of distinct PT specificities. One of 5 identical experiments is shown in figure 1. Analysis of supernatants with HPLC revealed two distinct PT specificities with 4.2 and 6.7 min. retention respectively, which were exclusively detected in supernatants of alloantigen-stimulated T cells.

The production of both PTs paralleled responder cell proliferation but preceded the appearance of alloantigen specific T killer cells by 3 to 4 days. It thus appeared that release of the 4.2 and 6.7 PTs is already a feature of alloantigen reactive lymphocytes during the recognitive phase of this response.

The second series of experiments aimed to unravel the genetic control of this alloantigen induced production of distinct PTs in vitro. T lymphocytes were stimulated in vitro with allogeneic cells differing from the responder cells in the degree of HLA disparity. One example, in which cells of two pairs of identical siblings differing from each other in only one HLA haplotype were studied, is depicted in Table 1.

FIGURE 1: Correlation between cellular proliferation and release of PT 4.2 and 6.7 into supernatants of human mixed lymphocyte cultures.

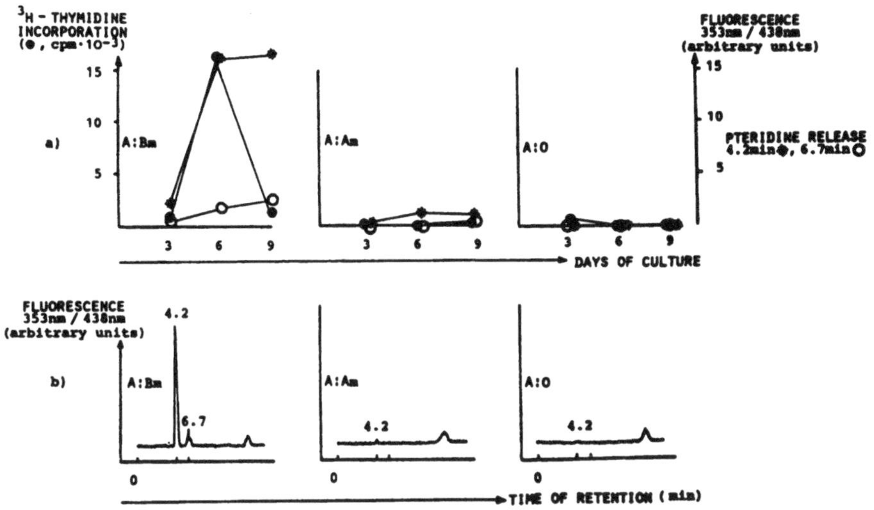

(a) Mixed lymphocyte cultures were established by co-cultivation of responder T cells from individual A with mitomycin-C treated stimulator cells from an unrelated individual B (A:Bm). Cultures of A cells with autologous and mitomycin-C treated cells (A:Am) and cultures of A cells (A:O) alone served as controls. At times indicated cells were assessed for proliferation by ^{3}H-TdR incorporation (●) and supernatants for production of PT 4.2 (*) and 6.7 (O) by HPLC.

(b) High pressure liquid chromatograms from supernatants of cultures on day 6. Values are corrected for fluorescence of control media incubated in the absence of cells.

In this, as well as in the other experiments, production of 4.2 and 6.7 PTs and proliferation of responder cells was exclusively a feature of T lymphocytes co-cultivated with HLA disparate stimulators. We thus conclude that the production of these two PT specificities in mixed lymphocyte culture is controlled by genes coded for by the HLA complex.

TABLE 1: Pteridine 4.2 and 6.7 release into supernatants of mixed lymphocyte cultures is genetically controlled by genes coded for by the HLA complex.

HLA-haplotype responder	HLA-haplotype stimulator	proliferation (cpm x 10^{-3})	pteridine production*** 4.2 min	pteridine production*** 6.7 min
M.J. (a/b)*	M.J. (a/b)	0.5	-**	-
	M.B. (a/c)	24.8	70.2	25.1
	M.H. (a/b)	0.3	-	-
	M.S. (a/c)	13.6	19.2	-
M.B. (a/c)	M.J. (a/b)	38.8	59.9	19.8
	M.B. (a/c)	0.2	-	-
	M.H. (a/b)	31.5	57.6	27.3
	M.S. (a/c)	0.5	-	-
M.H. (a/b)	M.J. (a/b)	0.7	-	-
	M.B. (a/c)	32.7	71.7	27.2
	M.H. (a/b)	1.0	-	-
	M.S. (a/c)	14.6	42.5	-
M.S. (a/c)	M.J. (a/b)	25.7	61.1	23.4
	M.B. (a/c)	1.2	-	-
	M.H. (a/b)	25.1	84.0	26.4
	M.S. (a/c)	0.3	-	-
M.J.	none	0.2	-	-
M.B.	none	0.6	-	-
M.H.	none	0.5	-	-
M.S.	none	0.5	-	-
none	M.J.	0.3	-	-
none	M.B.	0.4	-	-
none	M.H.	0.2	-	-
none	M.S.	0.2	-	-

* Peripheral blood mononuclear cells from four siblings (M.J., M.B., M.H., M.S.) were studied in one way mixed lymphocyte cultures as described in Fig. 1. On day 6 cells were assessed for proliferation by ^{3}H-TdR incorporation and supernatants for pteridine production by HPLC. The HLA-haplotype symbols are: a(A1,B8,DR2), b(A1,B17,DR5),c(A2,B13,DR3)

** not detectable

*** PT production/ml/10^6 cells for specificity 4.2 is given in arbitrary fluorescence units and for 6.7 in ng.

Discussion

In this paper we present evidence that T lymphocytes upon activation with histoincompatible HLA antigens will release large amounts of two distinct pteridines: one has already been identified as neopterin (2), the fine structure of the other is still under investigation. These two PTs were not detected in culture supernatants of T cells stimulated with autologous or histocompatible HLA identical cells. They were also absent or detectable only in trace amounts in supernatants of other mononuclear cells such as B lymphocytes, natural killer cells or monocytes/macrophages (results not shown). It thus appears that the release of these PTs is genetically controlled by genes coded for by the HLA complex and that the T lymphocyte is their main or the sole producer.

From the theoretical point of view these findings are of considerable importance: so far pteridines have been found to be associated with almost every form of life. Nevertheless, their general biological significance is still obscure. Here we report on a direct relation between activation of immune cells and PT release in vitro. Whether PTs produced by activated T cells are involved in immunoregulation or whether they represent metabolic by-products remains to be further elucidated.

From the practical point of view the production of specific marker procedures by certain activated T lymphocytes represents an ideal means for the biochemical assessment of such cells in vivo. T cells activated by foreign as well as by chemically or virally altered self-HLA antigens represent the crucial immune effector mechanism controlling self integrity. We feel that clinical evaluation of these cells by PT markers will gain insight into important aspects of pathological states: (a) into the participation of activated T cells on various diseases and (b) into possible correlations between the numbers of activated T cells and the clinical course. The subsequent contribution will provide an example for this application: allotransplant rejections are mediated by activated T cells and rejection episodes are accompanied and preceded by increased PT release.

References

1. Hausen, A., Fuchs, D., König, K., Wachter, H: J. Chromatogr. 227, 61-70 (1982)
2. Wachter, H., Hausen, A., Grassmayr, K. Hoppe-Seyler's Z. Physiol. Chem. 360, 1957-1960 (1979)
3. Zier, K.S., Huber, Ch., Braunsteiner, H: Europ. J. Immunol. 7, 366-370 (1977)

Acknowledgements

We thank Mrs. A.M. Födinger and Mrs. J. Pichl for technical assistance. This work was financially supported by the Austrian Funds "Zur Förderung der wissenschaftlichen Forschung", the "Jubiläumsfonds der Österr. Nationalbank" and the "Legerlotz-Stiftung".

PTERIDINES AS A NEW MARKER FOR THE PREDICTION OF HUMAN ALLOGRAFT REJECTION

R. Margreiter, M. Spielberger
Dept. of Surgery I, University of Innsbruck,
Anichstr. 35, A-6020 Innsbruck, Austria.

D. Fuchs, A. Hausen, G. Reibnegger, H. Wachter
Dept. of Medical Chemistry, University of Innsbruck,
Fritz-Pregl-Str. A6020 Innsbruck, Austria.

Ch. Huber
Dept. of Internal Medicine, University of Innsbruck
Anichstr. 35, A-6020 Innsbruck, Austria.

Introduction

Despite prophylactic immunosuppressive treatment the majority of human allograft recipients experience major rejection episodes, which carry a serious risk of graft loss, possibly intensified if immunosuppressive medication has been delayed due to uncertainties in establishing the clinical diagnosis. Rejection diagnosis so far is mainly based on blood chemistry, isotope renal scanning and biopsy. With the exception of the latter, pathological tests reflect advanced functional impairment due to severe graft damage. Other tests, such as beta-2-microglobulin-determination have not been proven to be of any clinical value in early diagnosis of graft rejection (4). This points to the need for reliable biochemical parameters for the early detection of allograft induced host immune reactions. Our search for such a biochemical marker was based on two experimentally well established findings: (a) acute rejection of allografts is preceded by rapid proliferation of allo-

Biochemical and Clinical Aspects of Pteridines · Volume 1
Cancer · Immunology · Metabolic Diseases

antigen reactive lymphocytes (1) and (b) in vitro and in vivo proliferation is invariably associated with increased production of certain pteridines (5-8). In vitro alloantigen activated T-cells produce neopterin and another pteridine specificity (rt = 4.2 min.) (9-10), up to now fine structure of which has not been identified. To test for the in vivo detection of alloantigen induced lymphocyte activation we have analyzed urine samples of human allograft recipients.

Patients and Methods

In 20 consecutive allogeneic cadaveric kidney and one liver transplant recipients the urinary neopterin excretion was daily assessed by high pressure liquid chromatography from the time after removal of the catheter, which was usually done on the third postoperative day, until discharge from the unit. Whereas only smaller amounts of the 4.2 min. retention pteridine were found in urine samples, neopterin was therefore chosen for clinical correlation. Patients' neopterin excretion in relation to urinary creatinine levels (neopterin-factor = µmol neopterin/mol creatinine) was compared with those of normal controls. The mean observation time was 23 (8-40) days. Donor and recipients were matched for HLA-A, -B and DR, respectively. Prior to transplantation all patients received at least three units of leucocyte poor blood. Posttransplant immunosuppressive therapy consisted of steroids and azathioprine or Cyclosporin A respectively. Diagnosis of rejection was based on clinical symptoms, blood and urine chemistry, fine needle aspiration biopsy and renal scanning. Rejections were treated with gram-doses of methylprednisolon up to a maximum of 6 grams on consecutive days.

Results

Based on neopterin excretion levels patients were divided

into two clearly defined groups: one presenting with normal and the other with increased levels.

The group with normal neopterin excretion throughout the time of observation comprised 33.3% of the patients. Their clinical course was uncomplicated and their graft function excellent (Fig. 1). In contrast, all patients with transitory or persistent elevation of neopterin excretion experienced major clinical complications; in 10 patients abnormal excretion was followed by one or more acute rejection episodes (R) (Fig. 2), which could be reversed in all but two cases (Fig. 3). Increased urinary neopterin levels preceded clinical diagnosis by at least 24-48 hours. In 3 recipients lacking evidence for rejection elevated neopterin excretion was associated with cytomegalovirus infection. In these three patients neopterin values reached significantly higher peaks than those seen in patients with acute graft rejection (Fig. 4). In one further case increased neopterin excretion due to rejection was followed by extremely high neopterin values, which were explained by an additional viral infection (Fig. 5).

Figure 1 Neopterin excretion of a patient with uncomplicated clinical course

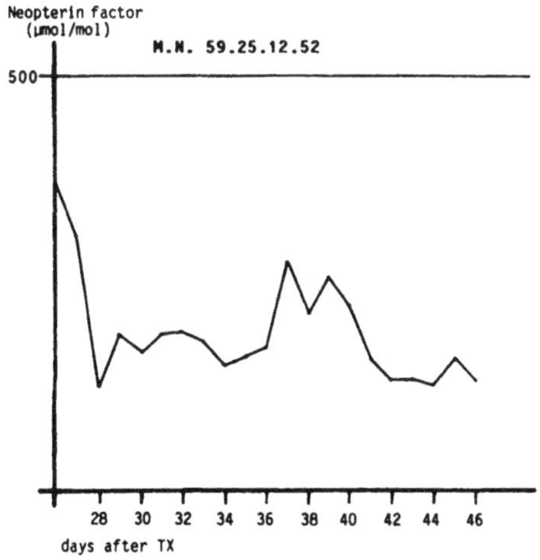

Figures 2,3

Neopterin excretion of two patients during acute rejection episodes (R)

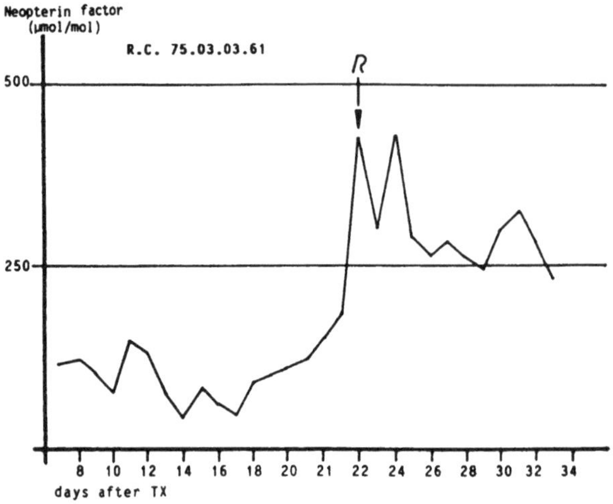

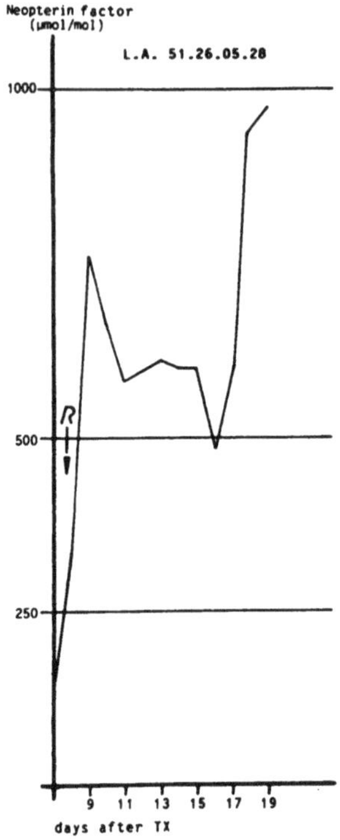

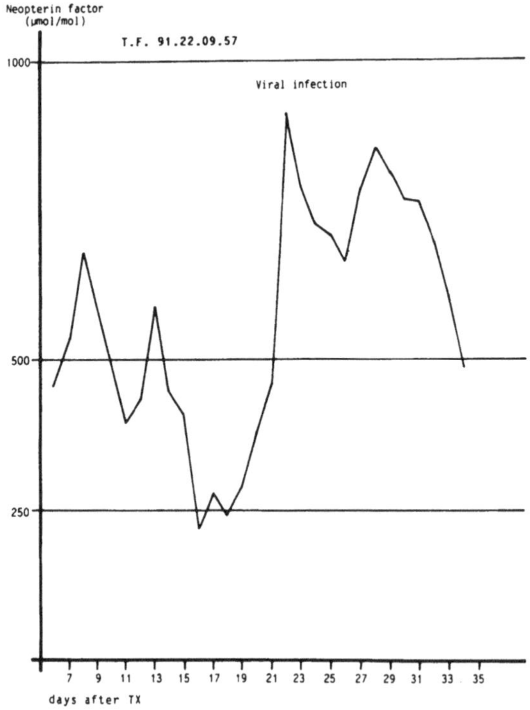

Figure 4

Neopterin excretion of a patient during a cytomegalovirus infection

Figure 5

Neopterin excretion of a patient during an acute rejection episode (R) followed by an additional viral infection (V)

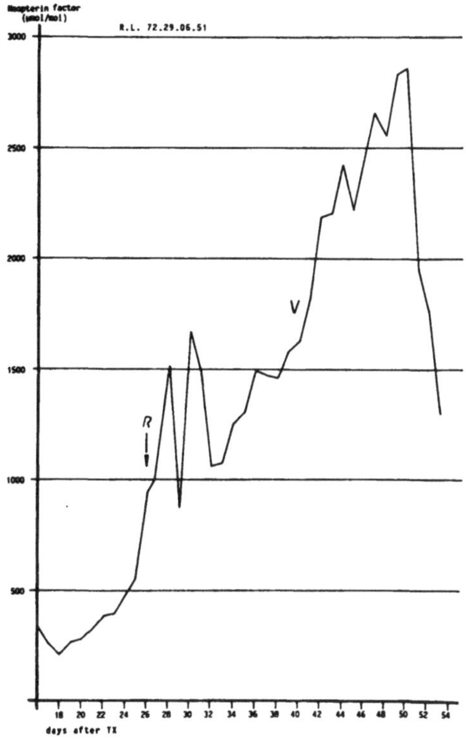

Figure 6

Neopterin excretion during a change of immunosuppressive therapy (↓)

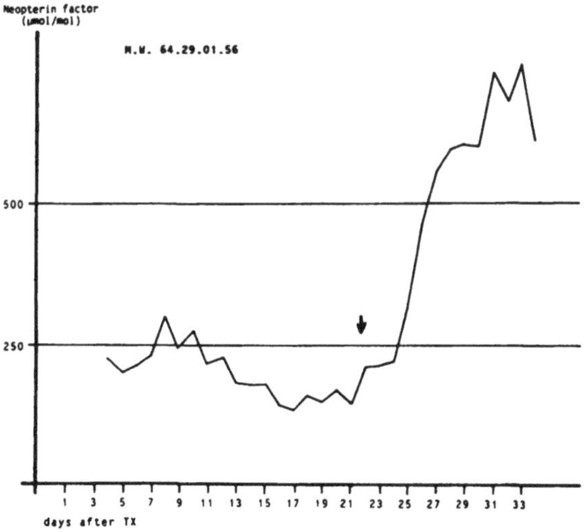

In one patient taking Cyclosporin A as the sole immunosupp-ressive agent this drug had to be withdrawn because of nephro-toxicity on day 21. A striking rise in PT excretion could be observed thereafter (Fig. 6).

Discussion

Mixed lymphocyte culture studies suggested that activation of T lymphocytes by allogeneic blood mononuclear cells leads to a specific production of two distinct PT specificities with retention times of 4.2 and 6.7 min respectively. This pro-duction paralleled T cell proliferation and was genetically controlled by genes of the major histocompatibility complex. Strikingly similar PT patterns were observed in vivo following allograft transplantation: increased amounts of neopterin (the 6.7 min PT) were excreted during every rejection episode, whereas normal levels were seen in patients without clinical evidence for immune mediated graft destruction. It thus ap-pears that in vitro as well as in vivo alloreactive lympho-cytes during their clonal expansion will produce large amounts of these PTs. This view is further supported by our observa-tion that cyclosporin A, a drug which is known to selectively interfere with T cell proliferation (2) effectively controlled PT release. Association of increased PT production with viral infections have previously been reported (7) and were also ob-served in the present study. This is not surprising because such infections may lead to modification of major histocompa-tibility antigens which then will be recognized as foreign by immune T cells and will subsequently be rejected (3). It therefore seems likely that increased amounts of PTs are not only produced during proliferation of alloantigen reactive lymphocytes but also during proliferation of immune cells re-cognizing modifications of"self"surface structures.

From the clinical point of view evaluation of PT excretion seems to represent a reliable marker for monitoring allograft recipients. So far increased levels in transplant recipients have been exclusively seen in association with rejection episodes or viral disease. In both situations an increase was seen at least one day before clinical diagnosis was established by standard procedures. Although increased PT levels are not specific for allograft rejection they enable identification of a group of patients with major complications which can be readily worked up to define the underlying disease process.

Acknowledgements

We thank Mrs. A.M. Födinger and Mrs. J. Pichl for technical assistance. This work was financially supported by the Austrian Funds "Zur Förderung der wissenschaftlichen Forschung", the "Jubiläumsfonds der Österr. Nationalbank" and the "Legerlotz-Stiftung".

References

1. Bach, F.H., Bach, L.M., Sondel, P.M.: Nature 259, 273-281 (1977).
2. Borel, J.F.: Transplant. Proc. 13, 344-348 (1981)
3. Doherty, P.C., Zinkernagel, R.M.: Lancet i, 1406-1409 (1975).
4. Roxe, D.M., Siddiqui, F., Santhanam, S., del Greco, F. Wolf, J.: Nephron. 27, 260-264 (1981).
5. Wachter, H., Hausen, A., Reider, E., Schweiger, M: Naturwiss. 67, 610-611 (1980).
6. Wachter, H., Grassmayr, K., Hausen, A.: Cancer Lett, 6, 61-66 (1979).
7. Wachter, H., Hausen, A. Grassmayr K.: Hoppe-Seyler's Z. Physiol. Chem. 360, 1957-1960 (1979).

8. Hausen, A., Fuchs, D., Grünewald, K., Huber, H., König, K., Wachter, H.: Clin. Chem. Acta 117, 297-305 (1981).

9. Fuchs, D., Hausen, A., Huber, Ch., Margreiter, R., Reibnegger, G., Spielberger, M. and Wachter, H.: Hoppe Seyler's Z. Physiol. Chem. 363, 661-664, (1982).

10. Huber, Ch., Fuchs, D., Hausen, A., Margreiter, R., Reibnegger, G., Spielberger, M. and Wachter, H. in preparation.

BLOOD LEVELS OF BIOPTERIN AND AGP_M DURING KIDNEY AND BONE MARROW TRANSPLANTATION

Michael Fink, Ulrich Jehn, Wolfgang Wilmanns
III. Medizinische Klinik der Universität, Klinikum Großhadern, D-8000 München 70, Germany

Irmgard Ziegler
Institut für Toxikologie und Biochemie der Gesellschaft für Strahlen- und Umweltforschung, Abteilung Zellchemie, D-8000 München 2, Germany

Introduction

In an earlier study, the biopterin blood level proved to be an indicator for enhanced cell proliferation (1). This may be of clinical significance, since elevated biopterin levels were found in the malignant cells during leukemias and Polycythemia vera (2). Furthermore, biopterin was an indicator of hemopoesis during experimental bone marrow transplantation (3). Therefore, it seemed to be possible, that the biopterin content of whole blood increases during allograft rejection after renal transplantation. This prompted us to compare biopterin levels with clinical data during renal graft rejection, which is indicated by pain of the transplanted kidney and by reduced diuresis. In comparing the blood biopterin levels with serum creatinine it has to be taken into account, that both substances are excreted in a similar fashion; in consequence, biopterin and creatinine levels are closely correlated (4; see Fig. 1). Therefore, an increase of biopterin correlated with creatinine can not be interpreted as an indicator for graft rejection, but only for the concomitant renal insufficiency. Only biopterin levels which exceed this correlated increase

Biochemical and Clinical Aspects of Pteridines · Volume 1
Cancer · Immunology · Metabolic Diseases

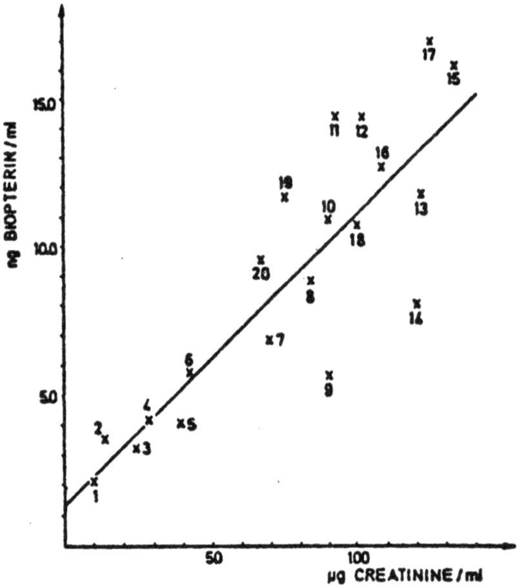

Fig.1 Correlation of biopterin and creatinine levels in whole blood during renal insufficiency. Patients No. 8-20 were subjected to dialysis. Correlation index=+0.86; slope=0.1.

can be considered as indicative for cell proliferation.

A pteridine binding variant of the alpha$_1$-acid-glycoprotein (AGP$_M$) was found to be an indicator of the acuteness of malignant and autoimmune diseases (6,7). This parameter was also examined in patients with kidney and bone-marrow transplants.

Material and methods

Venous blood was drawn into sodium citrate (1:10) and frozen at -15°C until use. Extraction, cleavage of protein bound biopterin by autoclaving with aconitate buffer at pH 4,5 and subsequent determination in the bioassay with Crithidia fasciculata has been described elsewhere (2,3,5).

AGP$_M$ was expressed in terms of protein bound tyrosine in "Fraction b". This fraction had been obtained from blood extract with digitonin followed by alcohol and TCA-precipitation

(6,7). Creatinine was determined by the method of Jaffe in the modification of Chasson et al. (8).

Results

1) Renal transplantation. The follow up of 16 patients after renal transplantation has been examined. Five examples are outlined in the following figures.

Uncomplicated renal transplantation is characterized by a slight overshoot of diuresis in the first few days after transplantation and a corresponding decrease of elevated creatinine levels. The biopterin levels decreased one or two days earlier than creatinine concentrations and did not exceed values correlating with creatinine accumulation (Fig.2).

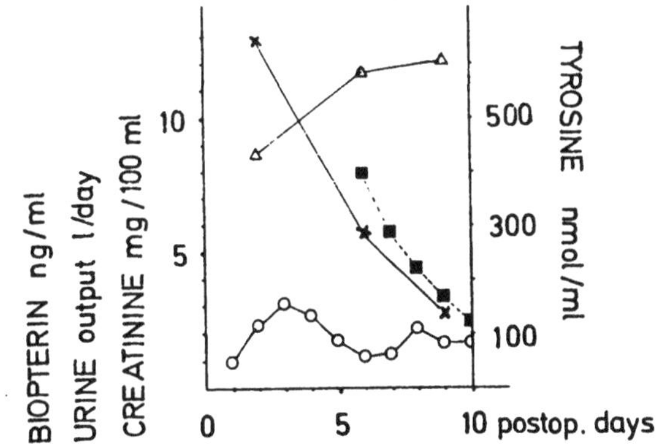

Fig.2. Biopterin and AGP_M-levels compared with urine output and creatinine after an uncomplicated renal transplantation; the biopterin decreases 1-2 days earlier than creatinine. Normal ranges: biopterin 3.0±0.66 ng/ml; AGP_M 280±80 nmol tyrosine/ml. AGP_M was indirectly determined by protein bound tyrosine in "Fraction b" (see ref.6,7). ×—× biopterin;△—△ tyrosine;o—o urine output; ■—■ creatinine;□—□ creatinine at dialysis.

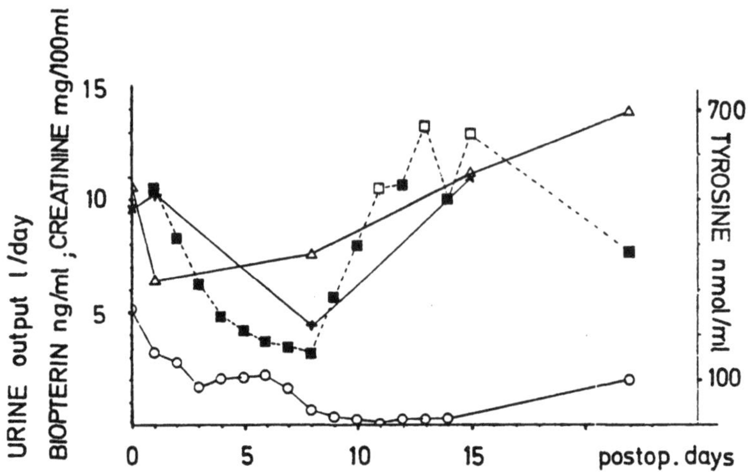

Fig.3. Biopterin and AGP_M-levels compared with urine output and creatinine in a patient with acute vasculo- and glomerulopathy and a moderate cellular rejection on the 8th day post transplant. On the day of rejection biopterin is slightly higher than creatinine. Symbols and normal values see Fig.2.

Pat. A.K. (Fig.3) showed an histologically proven acute vasculo- and glomerulopathy and a moderate cellular rejection on the 8th day post transplant. Biopterin data of the same day were found to be moderately higher than those correlating with increased creatinine. This rise of biopterin could possibly indicate a graft rejection earlier than the other parameters. Possibly a screening at shorter intervals could result in a diagnostic tool. During the following days, biopterin of patient A.K. increased in the usual correlation to creatinine and did not seem to be influenced by the treatment with anti-lymphocyte globulin, which was given from day 11-20.

In both cases, AGP_M was increased and did not seem to be correlated with the clinical course.

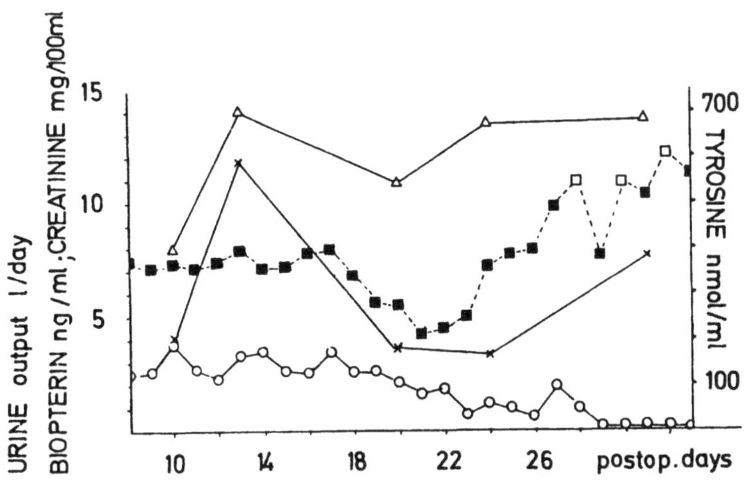

Fig.4. Biopterin and AGP_M-levels compared with urine output and creatinine in a patient with a cellular graft rejection on day 13 and a severe humoral graft rejection in the 5th week after transplantation. High biopterin peak in the cellular rejection and low biopterin compared with creatinine in the humoral rejection. Symbols see Fig.2.

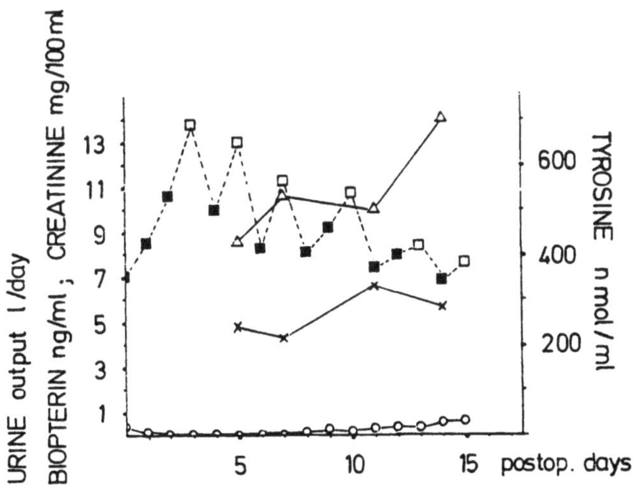

Fig.5. Biopterin and AGP_M-levels compared with urine output and creatinine in a patient with a cellular graft rejection on day 4. Low Biopterin at the time of cellular rejection. Anti-lymphocyte globulin was given from day 8-16. Symbols and normal values see Fig.2.

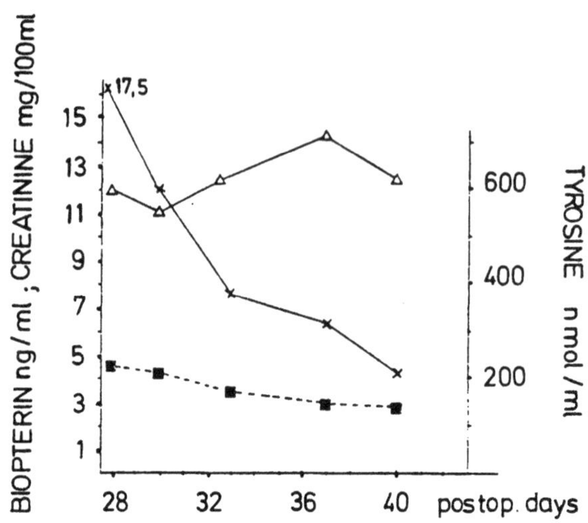

Fig.6. Biopterin and AGP_M-levels compared with creatinine in a patient with a HbSAg positive, anicteric hepatitis four weeks after kidney transplantation. The Biopterin value is highly elevated in relation to creatinine. Symbols and normal values see Fig.2.

Case A.H. (Fig.4) demonstrates a clear increase of biopterin above the correlating creatinine levels in a histologically proven cellular graft rejection on day 13, although - in the same patient - the humoral graft-rejection on day 30 was followed by an only slightly elevated biopterin level. However, case M.B. (Fig.5) could not further corroborate the view, that a cellular graft rejection is correlated with an elevated biopterin level.

In patient W.E. (Fig.6) highly elevated biopterin levels were found one month after transplantation. This was correlated with a HbSAg pos., anicteric Hepatitis (9), and not with renal disorders.

2) Bone marrow transplantation. Fig.7 demonstrates a bone marrow transplantation in a case of aplastic anemia. The patient, who had many transfusions before transplantation, received bone marrow from his blood group and HLA-identi-

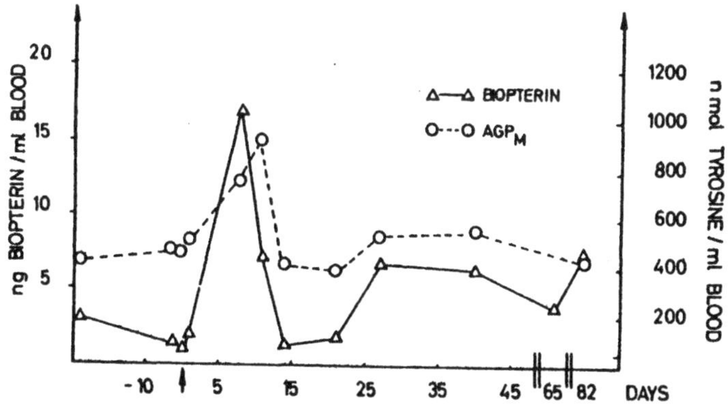

Fig.7. Biopterin and AGP_M in a patient with bone marrow transplantation; both values highly raised one week after transplantation. Normal values see Fig.2; details see text. ↑ day of transplantation

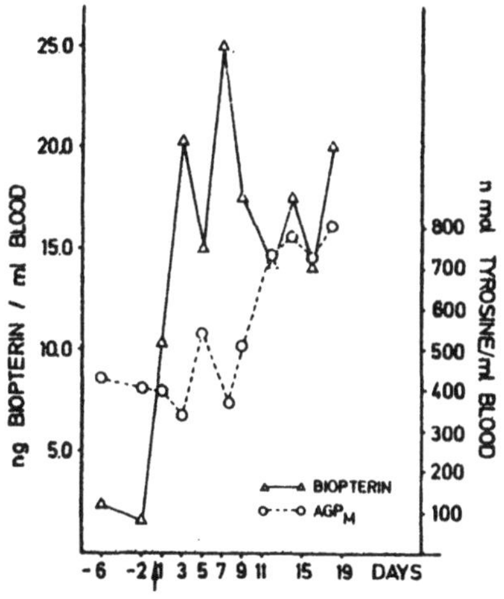

Fig.8. Biopterin and AGP_M in a patient with bone marrow transplantation; extremely high elevated biopterin levels after transpantation in spite of nearly normal renal function (creatinine < 2mg/100ml). Normal values see Fig.2; details see text. ↑ day of transplantation.

cal and MLC-negative brother. Before transplantation marrow was incubated with anti-T-cell rabbit globulin. After transplantation of the bone marrow, the patient developed chills and fever up to 40°C and suffered on the following day from severe edema in the mouth and pharynx. Fever continued during the following days and severe hepatomegaly of unknown etiology developed.. In a skin biopsy, a graft-versus-host disease could not been excluded completely. 26 days after bone marrow transplantation (BMT) the leucocytes rose over $1000/mm^3$, 54 days after BMT the reticulocytes rose up to 52‰; sufficient numbers of thrombocytes were never riched. The patient suffered from a cerebral bleeding and died of an interstitial pneumonia.

During the period of conditioning the biopterin levels dropped as they did in experimental BMT (3). On the 7th day after BMT, biopterin and AGP_M-levels exhibited high peaks in spite of a nearly normal renal function (Fig.7).

Another patient received a bone marrow transplantation from his blood-group and HLA-identical and MLC-negative brother during second remission of acute myelogenous leukemia. The course in the first six days after transplantation was inconspiciuous. Seven days after transplantation he developed fever of unknown origin and three days later, a severe erythrodermia developed. Fourteen days after BMT, the patient exhibited a progressive ikterus with ascites and hepatomegaly. He died 20 days after transplantation without signs of a take (Fig.8).

In spite of a nearly normal renal function with creatinine levels always under 2 mg/100ml, biopterin levels reached a peak of 25 ng/ml blood on day seven after BMT; this value was the highest ever found in our studies. During the following observation time, biopterin levels remained in a range of 15-20 ng/ml, while AGP_M increased up to 800 nmol tyrosine/ml blood.

Even though both clinical cases of BMT entailed highly increased levels of biopterin and AGP_M, at the moment a detailed interpretation would be premature.

Summary

Determinations of biopterin and AGP_M-levels in the whole blood after renal transplantation, so far seem to be of no prognostic value.

Two cases of bone marrow transplantation demonstrated clear changes in both biopterin and AGP_M-levels. As yet, the pathophysiological correlation is unexplained. The developement into a parameter of diagnostic value might be possible.

Acknowledgement

The kidney transplantations were performed in the Transplantations-Zentrum des Klinikums München-Großhadern; we thank Prof. Land for cooperation. The creatinine determinations were performed in the Institut für Klinische Chemie Großhadern (Director: Prof. M. Knedel).

References

1. Kokolis, N., Ziegler, I.: Zeitschr. f. Naturf. 23 b, 860-865 (1968).
2. Ziegler, I., Fink, M., Wilmanns, W.: Blut 44, 231-240, (1982).
3. Ziegler, I., Kolb, H.J., Bodenberger, U., Wilmanns, W.: Blut 44, 261-170 (1982).
4. Ziegler, I., Fink, M.: Biochem. Med. in press.
5. Baker, H., Frank, O., Shapiro, A., Hutner, S.H.. In: McCormick, D.B., Wright, L.D. (eds) Methods in Enzymology, vol 66. Academic Press, New York, 490-500 (1980).
6. Ziegler, I., Maier, K., Fink, M.: Cancer Res. 42, 1567-1573 (1982).

7. Fink, M., Ziegler, I., Maier, K., Wilmanns, W.: Cancer Res. 42, 1574-1578 (1982).

8. Chasson, A.L., Grady, H.J., Stanley, M.A.: Am. J. Clin. Path. 35, 83-88 (1961).

9. Wachter, H., Hausen, A., Grassmayr, K.: Hoppe Seyler's Z. Physiol. Chem. 360, 1957-1960 (1979).

5

PTERIDINES IN PROLIFERATION AND DIFFERENTIATION

PTERIDINES RELEASED FROM DICTYOSTELIUM DISCOIDEUM CELLS INTO THE EXTRACELLULAR MEDIUM

Günther Gerisch, Gertraud Blank
Max-Planck-Institut für Biochemie
D-8033 Martinsried bei München, Germany

Manfred Schweiger
Institut für Biochemie der Universität Innsbruck
A-6020 Innsbruck, Austria

Dietmar Fuchs, Arno Hausen, Gilbert Reibnegger, Helmut Wachter
Institut für Medizinische Chemie und Biochemie der Universität Innsbruck
A-6020 Innsbruck, Austria

The microorganism Dictyostelium discoideum is characterized by its unique developmental cycle: amoeboid cells grow by phagocytosis of bacteria or, under laboratory conditions, in liquid media. As long as nutrients are available, the cells remain single. In non-nutrient media the amoebae develop within several hours into cells which communicate with each other by cyclic AMP and other chemical signals, and they acquire the ability to form stable intercellular contacts. These cells aggregate into multicellular bodies using chemotaxis towards cyclic AMP and cell-to-cell adhesion as mechanisms of cell assembly.

Not only cyclic AMP but also folic acid has been shown to elicit a chemotactic response (1, 2). In contrast to cyclic AMP which acts most efficiently in the aggregation stage, folic acid acts as an attractant earlier in development. In the extracellular fluid folate is converted into 2-deamino folate. Folate deaminases responsible for this reaction are released from D.discoideum cells into the medium and are also present in their membranes (3, 4, 5, 6).

Biochemical and Clinical Aspects of Pteridines · Volume 1
Cancer · Immunology · Metabolic Diseases

Previous findings that a number of other non-identified pteridine derivatives are released from D.discoideum cells (2) raise the question of whether these compounds, or their precursors, play a regulatory role in development. Results obtained in other systems, including Ehrlich ascites tumor cells, Physarum

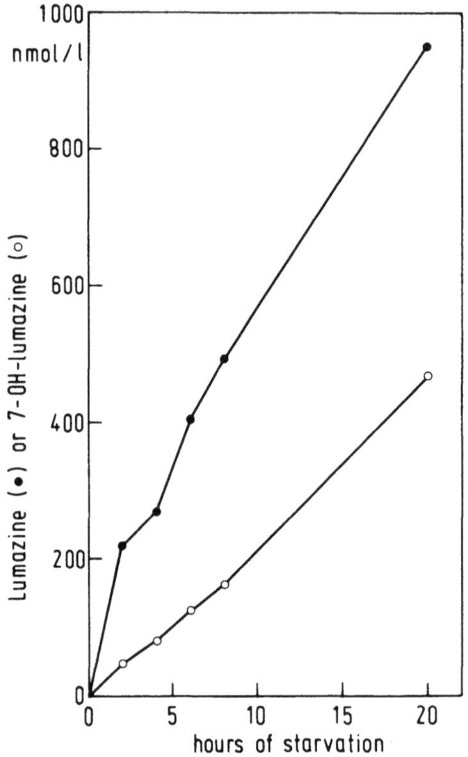

Fig. 1 Release of two pteridines from starved cells of D.discoideum, strain AX2-214. The cells were grown axenically as described (9), washed and resuspended at a density of $1x10^7$ cells per ml in 17 mM Soerensen phosphate buffer pH 6.0, and shaken at 23°C.
Abscissa: time after resuspension of washed cells in the buffer. Ordinate: concentrations of lumazine and 7-hydroxylumazine in the extracellular medium.

polycephalum and Acetabularia mediterranea indicate that the release of pteridines is related to cell growth (7). At least in some of these cases the pteridines are metabolites of GTP. This is evident for pteridines which carry a 3-carbon chain in position 6, like neopterin and monapterin, which is released from E.coli in a burst during the shift from exponential growth to the stationary phase (7).

The above mentioned results have prompted us to study the production and possible role of pteridines during development of D.discoideum. Five compounds have been separated by reverse phase HPLC as described (8). Identity of two of these products with lumazine and 7-hydroxy-lumazine, respectively, has been proven by their retention times on both a $C_{18}\mu$ Bondapak and a dimethylamino column, as well as by their fluorescence excitation and emission spectra. Both of these 2,4-dihydroxy pteridines are produced during growth in dark cultures. In non-growing cells their release is continued for at least 20 hours of starvation (fig. 1). The continued release of lumazine and 7-hydroxy-lumazine has also been observed in starved cells of a mutant, aggr 50, which is blocked early in development. Further work is required to clarify whether the pteridines released from Dictyostelium cells are waste products derived from physiologically important compounds like GTP or folic acid, or whether they play a direct role as intercellular messengers between growing or developing cells.

References

1. Pan, P., Hall, E.M., Bonner, J.T.: Nature New Biol. 237, 181-182 (1972).
2. Pan, P., Hall, E.M., Bonner, J.T.: J. Bacteriol. 122, 185-191 (1975).
3. Pan, P., Wurster, B.: J. Bacteriol. 136, 955-959 (1978).

4. Kakebeeke, P.I.J., De Witt, R.J.W., Konijn, T.M.: J. Bacteriol. 143, 307-312 (1980).

5. Bernstein, R.L., Rossier, C., Van Driel, R., Brunner, M., Gerisch, G.: Cell Diff. 10, 79-86 (1981).

6. Bernstein, R.L., Tabler, M., Vestweber, D., Van Driel, R.: Biochim. Biophys. Acta 677, 295-302 (1981).

7. Wachter, H., Hausen, A., Reider, E., Schweiger, M.: Naturwissenschaften 67, 610-611 (1980).

8. Hausen, A., Fuchs, D., König, K., Wachter, H.: J. Chromatogr. 227, 61-70 (1982).

9. Malchow, D., Nägele, B., Schwarz, H., Gerisch, G.: Eur. J. Biochem. 28, 136-142 (1972).

PTERIDINES DURING GROWTH AND DIFFERENTIATION OF PHYSARUM POLYCEPHALUM

Peter Loidl, Dietmar Fuchs, Peter Gröbner, Arno Hausen, Gilbert Reibnegger and Helmut Wachter

Institut für Medizinische Chemie und Biochemie der Universität, Fritz-Preglstraße 3/VI, A-6020 Innsbruck, Austria

Introduction

Recently pteridines have attracted considerable research interest since they seem to be implicated in the regulation of normal and malignant growth (1-5). The different pteridines, which have been detected in a variety of organisms and tissues, are all derived from GTP (6,7). The fact that GTP is degraded to pteridines, which are then excreted, suggests that these pteridines are regulatory molecules rather than waste products. To further elucidate the importance of pteridines during growth and differentiation we set out to study pteridines in the myxomycete Physarum polycephalum. This organism serves as a model system to study cell cycle dependent events, since nuclear division occurs naturally synchronous. Moreover there are several differentiation stages during the life cycle of Physarum, which make it possible to compare stages of proliferation with stages of quiescence (Fig.1). Excretion of isoxanthopterin as the main pteridine of Physarum has already been described (2).
The aim of this communication is to gain some insight into the regulatory function of pteridines. It could be demonstrated that pteridines are correlated with growth. A major function of pteridines could be a regulatory one during differentiation processes as shown for sclerotization of Physarum polycephalum.

Biochemical and Clinical Aspects of Pteridines · Volume 1
Cancer · Immunology · Metabolic Diseases

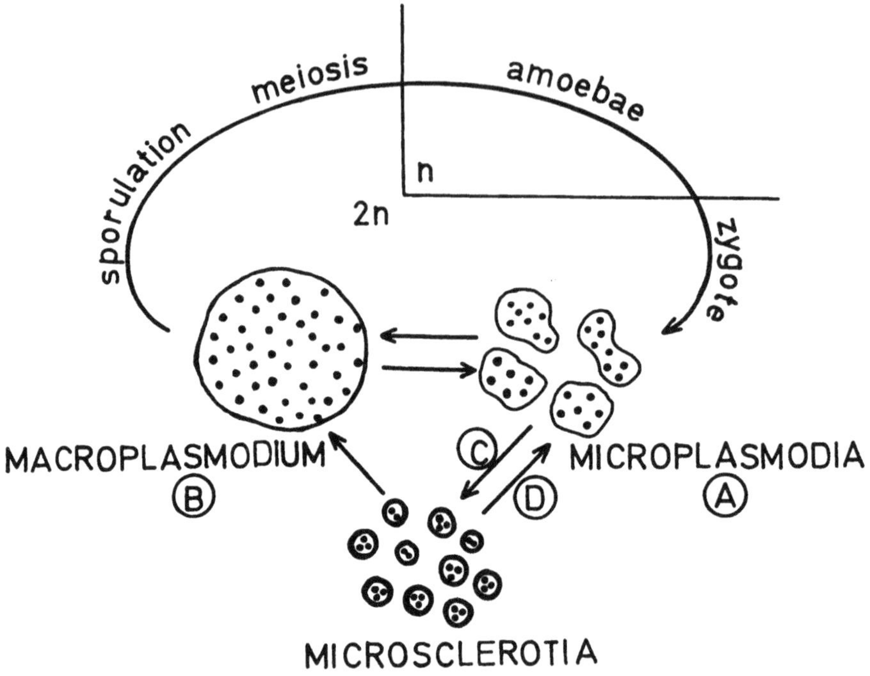

Fig.1. Schematic drawing of the life cycle of Physarum polycephalum.

Pteridines were studied in microplasmodia (A), macroplasmodia (B), during sclerotization of microplasmodia (C) and during germination of microsclerotia (D).

Materials and Methods

Chemicals and their sources were as follows: n-butyric acid from Sigma Chem.Co. (London, U.K.); FdUrd and Urd from Serva Feinbiochemica (Heidelberg, F.R.G.).

Microplasmodia of Physarum polycephalum (strain M_3b, a Wis1 isolate) were maintained under reciprocal shaking in submersed culture in semi-defined nutrient medium (8) supplemented with 0.013% hemoglobin instead of hematin. Macroplasmodia were prepared by fusion of microplasmodia and cultured in petridishes

on filter paper supported by glass beads. All culture techniques were carried out under sterile conditions.
The time of the synchronous mitosis was determined in ethanol-fixed smears of small explants of plasmodium under phase contrast (9). The intermitotic period of control plasmodia was 8 - 10 hours. For studies with metabolic inhibitors these were added to the growth medium.
Sclerotization of microplasmodia was induced as already described (8). Microsclerotia germination was done by suspending microsclerotia in 200 ml of sterile distilled water in a 500 ml Erlenmeyer flask and shaking for 30 min. at 24°C. The microsclerotia were then centrifuged, washed twice with distilled water and resuspended in several portions in one-third strength medium (8) in Erlenmeyer flasks, which were kept at 24°C under shaking.
For determination of isoxanthopterin content and excretion of microplasmodia, during sclerotization of microplasmodia and during germination of microsclerotia, replicate aliquots were removed from the culture and centrifuged. The supernatant culture fluid was used directly for pteridine determination, whereas for analysis of intracellular isoxanthopterin content the pellet was washed twice with distilled water and homogenized by sonication. The homogenate was centrifuged for 15 min. at 20000 x g at 2°C. The high-speed supernatant was used for pteridine determination.
For investigations during the synchronous nuclear division cycle of macroplasmodia 500 µl aliquots were removed from the culture medium for pteridine determination and replaced by fresh medium. Isoxanthopterin was determined by reversed phase high performance liquid chromatography as described for neopterin in detail (10). Fluorescence detector was set at an excitation wavelength of 341 nm and an emission wavelength of 408 nm, which corresponds to the maxima of isoxanthopterin in 15 mM phosphate-buffer (pH 6.4). 50 µl samples were injected for determination.

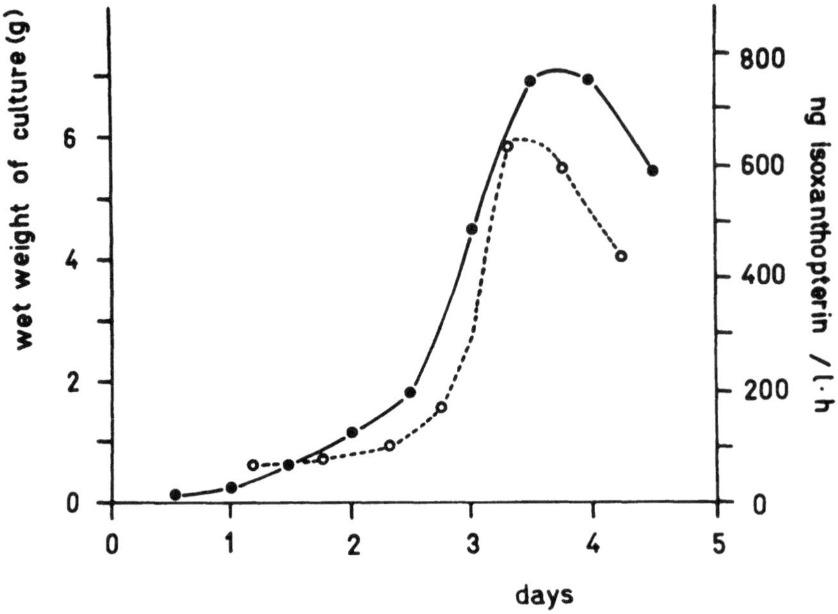

Fig.2. Excretion of isoxanthopterin during exponential growth of microplasmodia of Physarum polycephalum.

Replicate shake cultures of submersed microplasmodia were inoculated by transferring 0.1 ml plasmodial pellet into 20 ml of nutrient medium. At each timepoint one culture was harvested and the culture fluid was analyzed for isoxanthopterin as described under "Materials and Methods". The wet weight of the total plasmodial pellet was determined.
(●———●) wet weight (g); (o------o) excretion of isoxanthopterin (ng/l.h).

Results and Discussion

In exponentially growing microplasmodia the excretion of isoxanthopterin into the culture medium follows the growth rate of the culture. When microplasmodia reach the stationary phase isoxanthopterin excretion ceases (Fig.2). In order to clarify whether pteridine excretion is continuous or periodic during

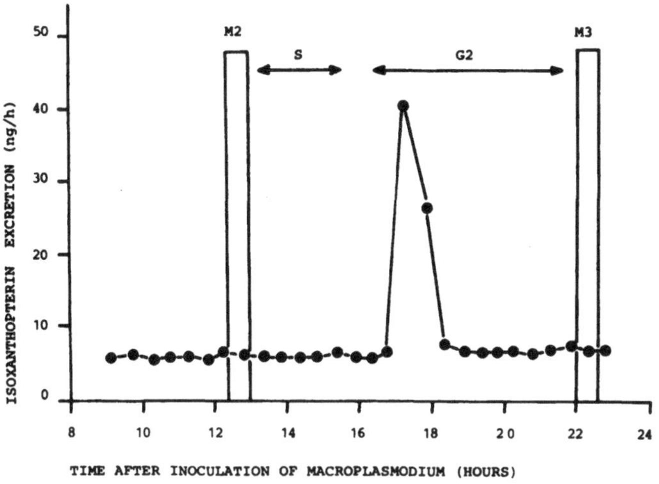

Fig.3. Excretion of isoxanthopterin during the synchronous nuclear division cycle of Physarum macroplasmodia.

A macroplasmodium was prepared as described under "Materials and Methods". 4 hours before mitosis 2 (M2) the plasmodium was transferred to fresh medium. At each timepoint 0.5 ml aliquots of the medium were removed for pteridine analysis and replaced by the same amount of fresh medium. S, G2 indicate the approximate duration of the S-phase, G2-period, respectively. There is no G1-period in macroplasmodia of Physarum.

the nuclear division cycle we measured isoxanthopterin levels in the growth medium of synchronous macroplasmodia. We found that the excretion rate stays fairly constant throughout the cycle, except for a short time in the early G2-period, where we observed a prominent peak in the excretion rate (Fig.3). It should be pointed out that after this maximum the rate of excretion is the same as in S-phase. This indicates that some biochemical event takes place in early G2-period which leads to an elevated excretion of isoxanthopterin.

After this observation during the normal, undisturbed mitotic cycle we performed experiments with FdUrd or Na-n-butyrate, which both lead to defined perturbations of the cell cycle.

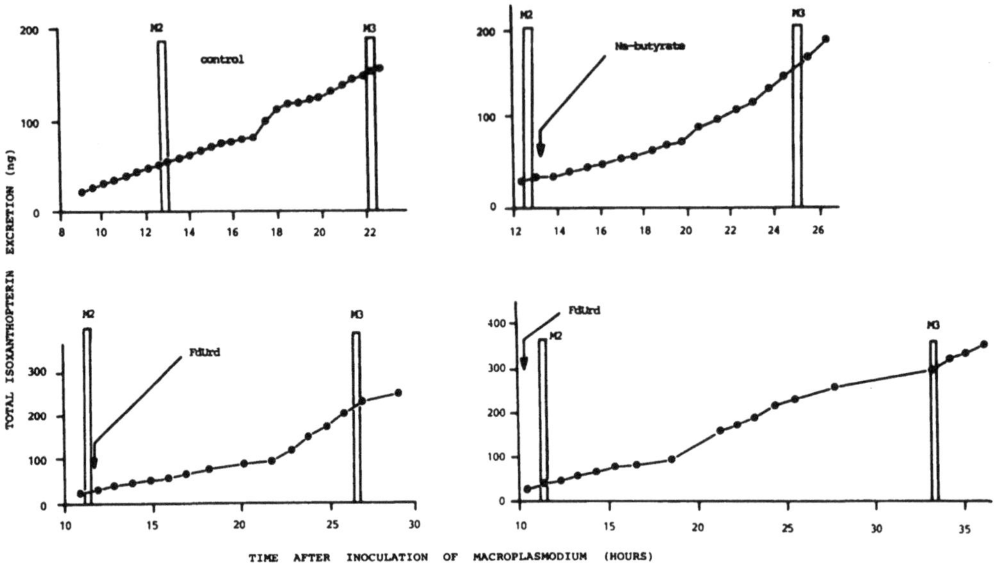

Fig.4. Isoxanthopterin excretion in macroplasmodia treated with FdUrd or Na-n-butyrate.

Macroplasmodia were prepared as described in "Materials and Methods". Plasmodia were transferred to fresh medium 30 min. before the first aliquot (0.5 ml) was analyzed for isoxanthopterin content. Na-n-butyrate (1 mM) or FdUrd (20 µM) + Urd (300 µM) were added at the times indicated. Urd was added to prevent incorporation of FUra into RNA (13). M indicates the time of mitosis.

FdUrd leads to a prolongation of the mitotic cycle by inhibition of DNA replication. Inhibition of DNA replication is caused by a strong inhibition of thymidylate synthetase (11), a key enzyme in dTMP biosynthesis. Na-n-butyrate inhibits DNA-, RNA- and protein - synthesis and therefore elongates the intermitotic period (12). After application of both drugs we could not observe dramatic changes in the pattern of isoxanthopterin excretion. Fig.4 shows that there is no pronounced

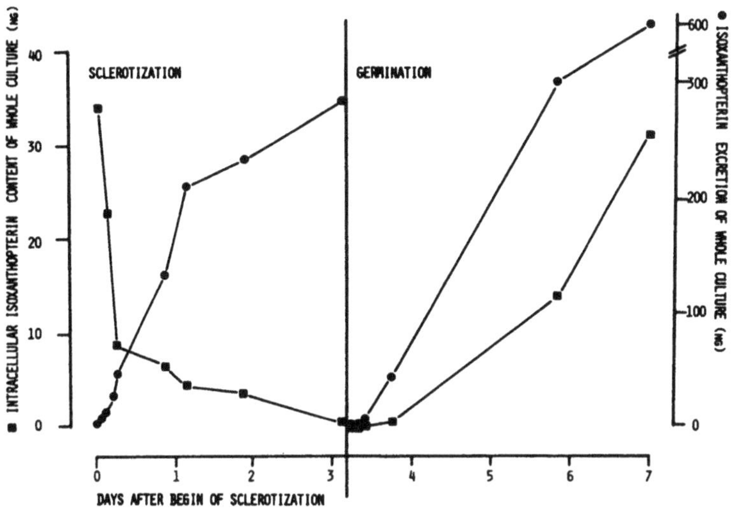

Fig.5. Isoxanthopterin excretion and intracellular isoxanthopterin content during differentiation (sclerotization). Sclerotization of microplasmodia and germination of microsclerotia was performed as described under "Materials and Methods". For determination of isoxanthopterin replicate cultures were harvested during the timecourse of the experiment and intracellular (■——■) as well as excreted (●——●) isoxanthopterin was determined.

maximum in isoxanthopterin excretion rate as compared to the control. After application of the compounds the excretion seems to be slightly inhibited for several hours. After this slight initial inhibition the level of excreted isoxanthopterin increases faster as compared to the control, which can be seen by the increased slope of the curves of treated plasmodia in Fig.4. It should be noted that the amount of isoxanthopterin excreted during one mitotic cycle is in strong correlation with the duration of the cycle. The pattern of isoxanthopterin excretion in the control and drug-treated plasmodia could in some way be associated with RNA- or protein-synthesis. This suggestion is in line with an overproduction of RNA and protein following butyrate treatment (12). In the case of FdUrd we also observed overproduction of RNA and protein in relation to DNA (Gröbner and Loidl, unpublished data).

To gain some understanding of the role of pteridines during differentiation we measured intracellular isoxanthopterin and isoxanthopterin excretion during sclerotization of microplasmodia. This differentiation process represents a transition from proliferation to quiescence. During this process we observed a dramatic release of isoxanthopterin into the medium with a concomittant decrease of intracellular pteridine. At the end of sclerotization there was only a hardly detectable isoxanthopterin level in microsclerotia (Fig.5). In contrast we found a ten fold amount of isoxanthopterin in the medium of microsclerotia as compared to the initial isoxanthopterin content at the beginning of sclerotization. When microsclerotia were germinated to form growing microplasmodia -a transition from quiescence to proliferation- we observed an increase in pteridine concentrations, both in the medium and intracellularly. Strikingly pteridines were detected in the medium before any pteridine could be found intracellularly (Fig.5). This could be explained either by a quantitative release of newly synthesized isoxanthopterin into the medium at early stages of germination or by the association of the relevant enzymes for the production of pteridines from GTP as precursor with the membrane. In Dictyostelium one of the relevant enzymes, a pterin deaminase, is located on the outside of the membrane and has already been characterized (14-16).

We have shown that pteridines are excreted during proliferation and differentiation of Physarum polycephalum. The fact that a burst of pteridine excretion was observed during differentiation of microplasmodia into microsclerotia leads to the suggestion, that pteridines have an important regulatory function during induction of differentiation processes. It should be pointed out that microplasmodia immediately stop growth, when they are induced to form microsclerotia; the fact that pteridines are excreted into the medium at a strongly increased rate during this differentiation step could be interpreted as a differentiation specific signal. The molecular mechanism of this signal, however, remains to be characterized.

References

1. Wachter, H., Grassmayr, K., Hausen, A.: Cancer Letters 6, 61-66 (1979).
2. Wachter, H., Hausen, A., Reider, E., Schweiger, M.: Naturw. 67, 610-611 (1980).
3. Hausen, A., Fuchs, D., Grünewald, K., Huber, H., König, K., Wachter, H.: Clin. Chim. Acta 117, 297-305 (1981).
4. Rokos, H., Rokos, K., Frisius, H., Kirstaedter, H.J.: Clin. Chim. Acta 105, 275-286 (1980).
5. Bichler, A., Fuchs, D., Hausen, A., Hetzel, H., König, K., Wachter, H.: Clin. Biochem. 15, 38-40 (1982).
6. Fukshima, K., Eto, J., Mayumi, T., Richter, W., Goodson, S., Shiota, T.: In Chemistry and Biology of Pteridines (Pfleiderer, W., ed.) pp. 247-261, Walter de Gruyter, Berlin · New York 1975.
7. Brown, G.M., Yin, J., Suzuki, Y., Heine, M.C., Foor, F.: In Chemistry and Biology of Pteridines (Pfleiderer, W., ed.) pp. 219-246, Walter de Gruyter, Berlin · New York 1975.
8. Daniel, J.W., Baldwin, H.H.: In Methods in Cell Physiology (Prescott, D.M., ed.) pp.9-41, Academic Press, New York 1964.
9. Guttes, E., Guttes, S., Rusch, H.P.: Dev. Biol. 3, 588-614 (1961).
10. Hausen, A., Fuchs, D., König, K., Wachter, H.: J. Chrom. 227, 61-70 (1982).
11. Gröbner, P., Loidl, P.: FEBS-Lett. 140, 41-44 (1982).
12. Loidl, P., Gröbner, P., Csordas, A., Puschendorf, B.: In Biology of Physarum (Rakoczy, L., ed.) pp. 90-97, Jagiellonian University Press, Cracow 1981.

13. Birnie, G.D., Kröger, H., Heidelberger, C.: Biochemistry 2, 566-572 (1963).

14. Pan, P., Wurster, B.: J. Bacteriol. 136, 955-959 (1978).

15. Darmon, M., Brachet, P.: In Receptors and Recognition (Hazelbauer, G.L., ed.) series B, vol.5, pp.101-132, Chapman and Hall, London 1979.

16. Wurster, B., Bek, F., Butz, U.: J. Bacteriol. 148, 183-192 (1981).

BIOPTERIN IN PERIPHERAL BLOOD CELLS AS AN INDICATOR FOR TRANSPLANTATION INDUCED HEMOPOIESIS IN BEAGLE DOGS AND FOR LEUKEMIC PROLIFERATION

Irmgard Ziegler

Institut für Toxikologie und Biochemie der Gesellschaft für Strahlen-und Umweltforschung, Abteilung Zellchemie; D-8 München 2, Germany

Michael Fink and Wolfgang Wilmanns

III.Medizinische Klinik der Universität, Klinikum Großhadern; D-8 München 70, Germany

Hans-Jochen Kolb and Ulrich Bodenberger

Institut für Hämatologie der Gesellschaft für Strahlen-und Umweltforschung, Abteilung Immunologie und Abteilung Klinische Hämatologie; D-8 München 2, Germany

Introduction

During regeneration experiments with the newt, Triturus cristatus the marked ability of amphibians to form pterin revealed for the first time that cell proliferation was linked to biopterin synthesis (1). The following investigations into pterin synthesis accompanying human cell proliferation focused on pterin levels in blood (2) and urine (3,4,5).

In mammals, normal and especially transplantation induced hemopoiesis may offer a system for studying stem cell proliferation, subsequently followed by differentiation; the products finally are found in the cellular fractions of the peripheral blood. Moreover, pterin concentration in those cells may offer a tool for detection of abnormal hemopoietic cell proliferation, e.g. during leukemias or during Polycythemia vera.

Biochemical and Clinical Aspects of Pteridines · Volume 1
Cancer · Immunology · Metabolic Diseases

Materials and Methods

The general conditions of fractionated total body irradiation, bone marrow cryopreservation and infusion in Beagle dogs are described in (6,7). Detailed conditions, the methods of blood cell separation and of biopterin determination by means of Crithidia assay are described in previous papers (8,9).

Results and Discussion

1) Biopterin during transplantation induced hemopoiesis in Beagle dogs. In Beagle dogs bone marrow aplasia was achieved by fractionated total body irradiation. Afterwards, hemopoiesis was started by infusion of autologous, cryopreserved bone marrow. During both these periods, biopterin was determined in the plasma, the red cells and the leukocytes from a constant volume of blood (1 ml). It was compared with the number of cells and the amount of cellular protein. The experiments, which are described in detail in (8) permit the following conclusions.

In adult Beagle dogs, the cellular portion of 1 ml blood averages 85 - 100 ug biopterin, of which only 10 ug are contained in the plasma fraction.

The marked decrease during the 7 day period of aplasia as well as the increase during blood cell restoration after autologous bone marrow infusion takes place predominantly in the cellular fraction (fig.1).

Within the cellular fraction, buffy coat biopterin largely correlated with the number of leukocytes reappearing in the peripheral blood (fig.2).

The red cell biopterin proved to be independent of the hematokrit, which did not change during the period of aplasia and of restitution (fig.3). It correlated with the

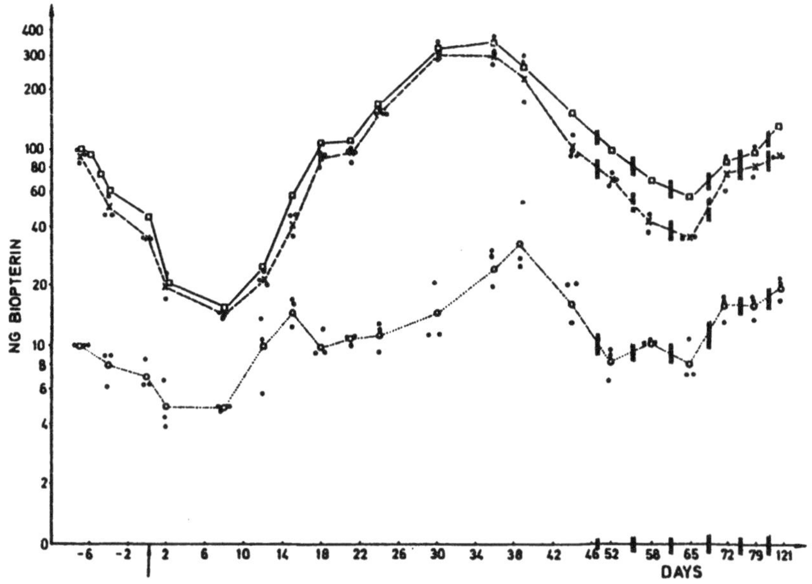

Fig.1 Biopterin levels in whole blood of dog N 651, in the corpusculate fraction and the plasma during whole body irradiation (days -7 to 0) and after infusion of autologous bone marrow. Each of the both dogs followed for both fractions showed the same type of curve. □——□ biopterin in whole blood (1 ml); x------x biopterin in the corpusculate fraction of 1 ml blood; o····o biopterin in the plasma portion of 1 ml blood. Arithmetic means of 2-4 determinations indicated by dots

number of reticulocytes. This indicates that during reticulocyte maturation, most of their biopterin is lost and only a small portion is retained in the mature erythrocytes. Consequently, during the period of increased reticulocyte invasion (days 23 - 35) the red cell fraction contains 1.3 ng biopterin /mg erythrocytic protein, compared to normal levels of 0.53 ng/mg protein.

A transient peak of plasma biopterin at day 38 constantly followed the period of overshoot in the reticulocytic biopterin. This corroborates the view that during maturation, most of the reticulocytic biopterin diffuses out of the cells and is transported by the plasma to be excreted.

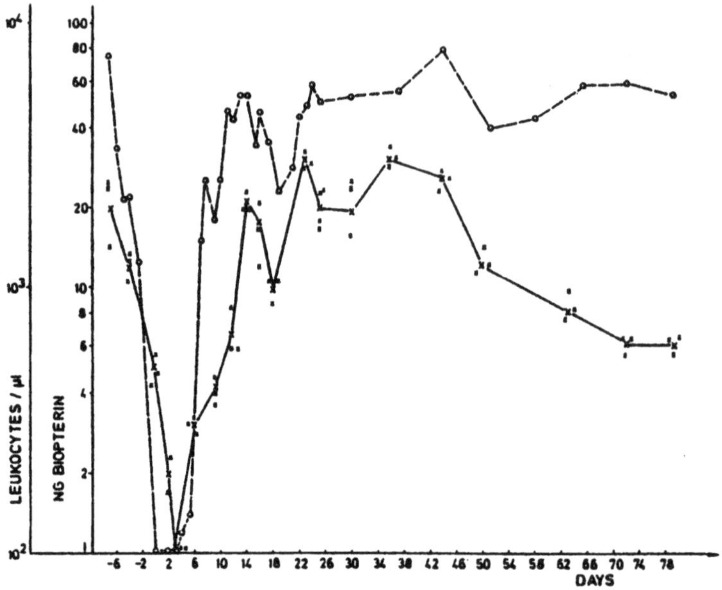

Fig.2 Biopterin levels in the buffy coat and number of leukocytes in the dog N 650 during whole body irradiation (days -7 to 0) and after infusion of autologous bone marrow x——x biopterin in the buffy coat of 1 ml blood; o------o number of leukocytes/ul blood; ↑ bone marrow infusion

In conclusion, the experiments demonstrate that biopterin is synthesized during hemopoietic cell proliferation and taken along with the cells into the peripheral blood. In the case of red cell maturation, going on in the peripheral blood, it is largely lost.

2) Biopterin during cases of leukemias and Polycythemia vera *. The normal distribution of biopterin among the various fractions of blood is given in table 1. Due to its predominant mass, the erythrocyte fraction comprises the bulk of cellular biopterin per volume of blood, whereas the lymphocytic contribution is much smaller. Calcula-

---

* CLL = chronic lymphocytic leukemia; ALL = acute lymphocytic leukemia; CML = chronic myelocytic leukemia; AML = acute myelocytic leukemia

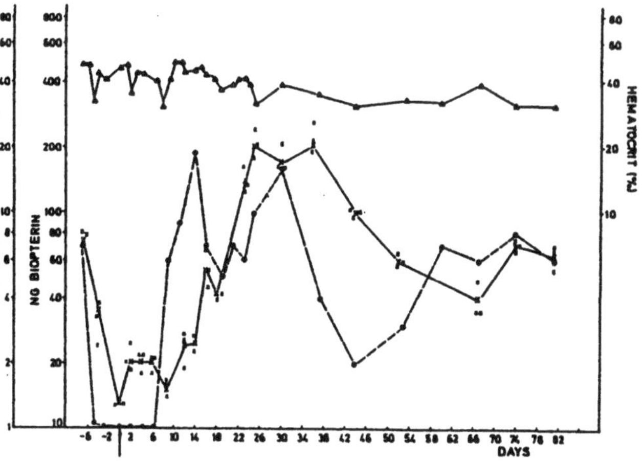

Fig.3 Biopterin levels in the erythrocyte fraction, number of reticulocytes and hematocrit in dog N 650 during whole body irradiation (days -7 to 0) and after infusion of autologous bone marrow. x——x biopterin in the erythrocyte fraction of 1 ml blood; o---o corrected reticulocyte count (o/oo); Δ——Δ hematocrit (%); ↑ bone marrow infusion

Table 1 Normal distribution of biopterin among the various fractions of blood (8 healthy donors). Volumetric indications refer to 1 ml blood. ± (S.D.)

Total	Plasma	Erythrocyte fraction		Buffy coat		Lymphocyte layer (separate determination)	
ng/ml	ng /ml	ng/ml	ng/mg protein	ng/ml	ng/mg protein	ng/ml	ng/mg protein
3.01	0.98	1.7	0.010	0.33	0.092	0.050	0.355
(0.53)	(0.31)	(0.18)	(0.0008)	(0.08)	(0.008)	(0.01)	(0.011)

ted on a protein basis however, the biopterin concentration is highest in the lymphocytes, the only blood cells (besides monocytes) which retain mitotic ability.

During normal renal function, plasma biopterin is readily excreted but is increased during renal insufficiency (10). Therefore, during most cases of leukemias, a clear cut increase in the plasma portion was only found if increased creatinine levels indicated a concomitant impairment of renal excretion (e.g. case H.F. in fig.6).

Leukemic disoreders are best manifested by changes in cellular biopterin. During all cases of chronic myelocytic leukemia, the amounts of buffy coat biopterin and of red cell biopterin were markedly increased (fig.4). The close correlation between CML and Polycythemia vera was demonstrated by a common increase in erythrocytic biopterin. The special accentuation of the red cell lineage during Polycythemia vera was manifested by maximal increases of erythrocytic biopterin in this disease (fig.5).

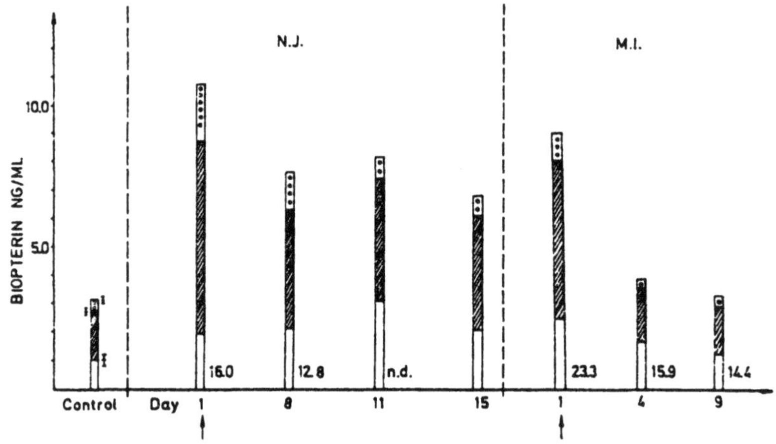

Fig.4 Biopterin levels of various blood fractions during blast crisis of CML. The fractions are isolated from 1 ml blood. ↑ cytostatic therapy; ▯ plasma; the adjacent numbers designate the amount of creatinine (ug/ml); ▨ erythrocyte fraction; buffy coat. Patient N.J.: died on day 21; Patient M.I.: remission

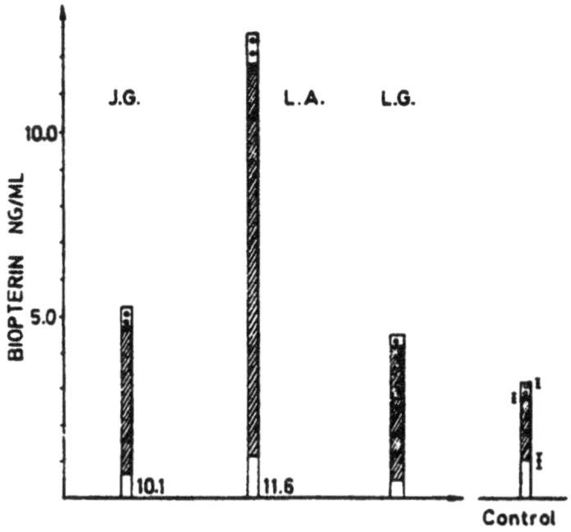

Fig.5 Biopterin levels of various blood fractions during Polycythemia vera. For explanation of symbols see fig.4

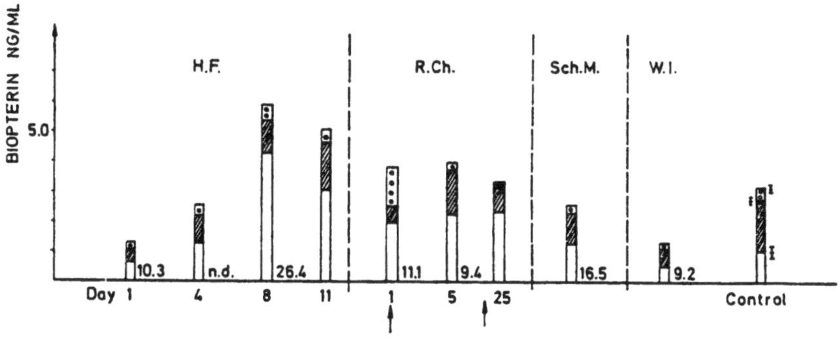

Fig.6 Biopterin levels of various blood fractions during CLL (case H.F.), ALL (T-ALL) (case R.Ch.), acute leukemia of mixed cellularity (lympho-and monocytic, aleukemic)(case Sch.M.) and ALL (aleukemic) (case W.I.).For explanation of symbols see fig.4

During all cases of ALL and CLL, the partial extinction of the red cell line was indicated by a marked decrease in the erythrocytic biopterin (fig.6). Due to the predominant mass of the erythrocytes and the low mass contribution of lymphocytes per volume of blood, lymphocytic leukemias are therefore characterized by a decrease in whole blood biopterin; whereas the increase in the myeloic cell lineage during CML and during Polycythemia vera is readily expressed by whole blood biopterin determinations (table 2).

Table 2 Biopterin levels in whole blood during leukemias and Polycythemia vera. Different samples were taken in at least 2 days intervals. Only those cases were taken during which normal creatinine values indicates normal renal function

	Mean ng/ml	± S.D. ng/ml	number of individuals	number of samples
Controls	3.0	0.66	65	65
CML (blast crisis)	8.0	4.40	7	23
AML (relapse)	8.0	5.40	3	24
CLL (relapse)	2.8	1.12	5	14
ALL (relapse)	4.2	2.11	4	8
Polycythemia vera	8.1	3.01	4	8

The myeloblasts, as present during CML and the lymphoblasts as found during ALL contribute high amounts of biopterin to a unit volume of blood. An even more characteristic feature of those blasts is their marked increase in biopterin concentration per unit of protein, as compared to the respective mature cells (figs.7,8) This is lacked by the lymphocytes during CLL and their mass accumulation only resulted in high amounts per blood volume (fig.8). The biopterin data thus agree with the current view on the pathophysiology of this leukemia as a predominantly accumulative disease.

In all of the 5 longitudinal sections,relapses and remissions could be monitored by following the biopterin levels in the peripheral blood cells. This is demonstrated in fig.4 by 2 cases of CML, which had opposite clinical results.

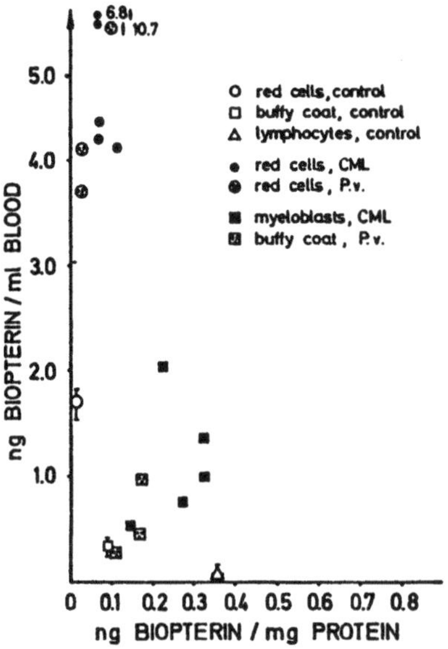

Fig.7 Amounts of biopterin per blood volume and concentration per unit of protein during CML and Polycythemia vera

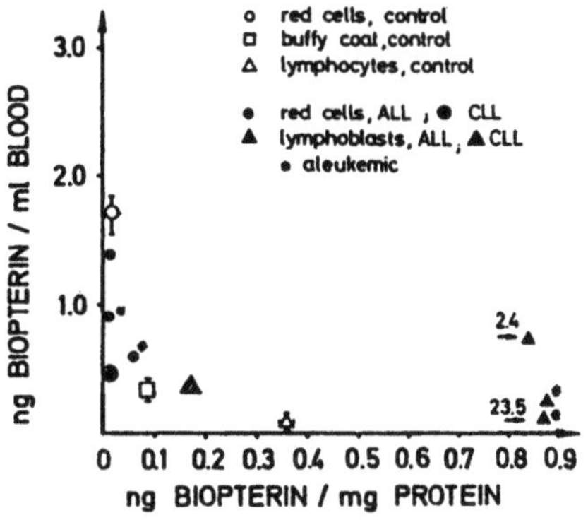

Fig.8 Amounts of biopterin per blood volume and concentration per unit of protein during CLL and ALL

The investigations of leukemias and of Polycythemia vera corroborate the view that there is an essential connection between biopterin synthesis and hemopoietic cell proliferation. The results agree with the current view on leukemogenesis (12). Therefore, a rapid assay for biopterin which is as sensitive as the Crithidia assay may provide a criterion for the presence of blasts especially in cases of aleukemic leukemias.

References

1. Kokolis,N., Ziegler,I.: Z.Naturf. 23b, 860 - 865 (1968)

2. Kokolis,N., Ziegler,I.: Cancer Biochem.Biophys.2,79-85 (1977)

3. Halpern,R., Halpern,B.C., Stea,B., Dunlap,A., Conklin,K., Clark,B., Ashe,H., Sperling,L., Halpern,I.A., Hardy,D., Smith,R.A.: Proc.Natl.Acad.Sci. (USA) 74, 587-591 (1977)

4. Rokos,H., Rokos,K., Frisius,H., Kirstaedter,H.I.: Clin. Chim.Acta 105, 275 - 286 (1980)

5. Hausen,A., Fuchs,D., Grünwald,K., Huber,H., König,K., Wachter,H.: Clin.Chim.Acta 117, 297 - 305 (1981)

6. Bodenberger,U., Kolb,H.J., Rieder,I., Netzel,B., Schaffer, E., Kolb, H., Thierfelder,S.: Exp.Hematol.8, 384-394 (1980)

7. Kolb,H.J., Rieder,I., Bodenberger,U., Netzel,B., Schaffer, E., Kolb,H., Thierfelder,S.: Pathol.Biol.27, 370-372 (1979)

8. Ziegler,I., Kolb,H.J., Bodenberger,U., Wilmanns,W.: Blut 44, 261-270 (1982)

9. Ziegler,I., Fink,M., Wilmanns,W.: Blut 44, 231-240 (1982)

10. Ziegler,I., Fink,M.: Biochem.Med., in press (1982)

11. Dameshek,W.: Blood 29, Suppl., 566-584 (1967)

12. McCulloch,E.A.: J.Natl.Cancer Inst.63, 883-892 (1979)

IMPORTANCE OF PTERIDINES FOR REGULATION OF GROWTH

H. Wachter, D. Fuchs, A. Hausen, G. Reibnegger

Institute for Medical Chemistry and Biochemistry, University of Innsbruck, Fritz-Pregl-Strasse 3, A-6020 Innsbruck, Austria

Previous investigations and results presented in this workshop demonstrate raised urinary neopterin excretion in patients with various malignancies (1), with viral diseases (2) and with allograft rejections (3). Despite great diversity one fact is common to all these diseases: activation of T-lymphocytes due to stimulation by tumor specific transplantation antigens, virus antigens or by alloantigens.

Table 1: Neopterin excretion due to cell-mediated immune response

Neopterin excretion of patients suffering from:	n	$\bar{x}$ µmol neopterin / mol creatinine	range µmol neopterin / mol creatinine
a) Intracellular microorganisms:			
Mycobacterium tuberculosis	20	379	225 - 884
Toxoplasma	1	11663	
Mycoplasma	1	1435	
b) Virus diseases (cytomegaly, epidemic parotitis, influenza, measles, rubella, varicella, catarrhal angina)	50	1286	302 - 4044
c) Cell-mediated hypersensitivity (contact dermatitis, surface involvement > 20cm^2)	10	262	153 - 536

Our idea is that elevated neopterin excretion is a result of activation of T-lymphocytes, which react against foreign surface structures to keep the self integrity of the organism.

Biochemical and Clinical Aspects of Pteridines · Volume 1
Cancer · Immunology · Metabolic Diseases

In this case, all processes in mammalians which evoke cell-mediated immune response, should lead to elevated urinary neopterin levels. In table 1 data are shown obtained from persons, who were suffering from diseases, in which cell-mediated immune responses are well known, and indeed raised neopterin excretion was found.

To verify all these observations we investigated cultured T-lymphocytes during proliferation and differentiation. The neopterin excretion into culture media found during allogeneic activation seems to be highly specific for T-lymphocytes during differentiation to cytotoxic effector cells. Neither other haematopoetic cell types nor many other cells studied were able to release neopterin to an appreciable extent (4).

The existing parallel between neopterin release and proliferation and differentiation, respectively, of T-lymphocytes, the release of isoxanthopterin by Physarum polycephalum during differentiation from microplasmodia to microsclerotia (5) and the release of different pterins and lumazines by Dictyostelium discoideum (6) suggests that the pteridines found were not incidental excretion products; moreover, pteridines, or especially their reduced forms, may have an important function in cellular metabolism besides the well established cofactor function of tetrahydrobiopterin.

To attempt preliminary hypotheses, concerning the action of pteridines in regulation of proliferation and differentiation, seems an interesting task. The following hypotheses should be regarded as a first step leading to further discussion:

The highest release of pteridines in culture media was always seen, when cells were just on a switch during their development (5,7) e.g. when they ceased to proliferate or started to differentiate. One might interpret this observation in that way that pteridines may regulate growth by activation or inactivation of inhibitory factors, such as repressors or chalones, in accordance with well known theories.

Another hypothesis is that the molecular action of pteridines for the regulation of development is considered to mask m-RNA sections. Formation of hydrogen bonds between pteridines and the bases of nucleotide moieties of m-RNA seems possible (Fig.1).

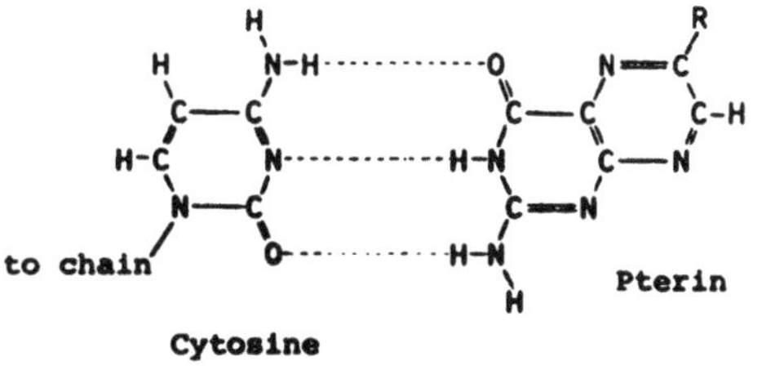

Fig.1: Hydrogen - bonds between a pteridine and a nucleoside

The m-RNAs masked in this way could not be translated into proteins at the ribosomes. Decrease of pteridine concentration results in decline of m-RNA masking. The pteridine levels, on the other hand, are diminished through increase of cellular membrane permeability by factors such as histamine or hormonelike substances. The incidence of high histamine levels in rapidly proliferating tissues is well known (8). The enhanced cellular membrane permeability lowers the intracellular pteridine concentration and finally terminates m-RNA masking. The regulatory role of pteri-

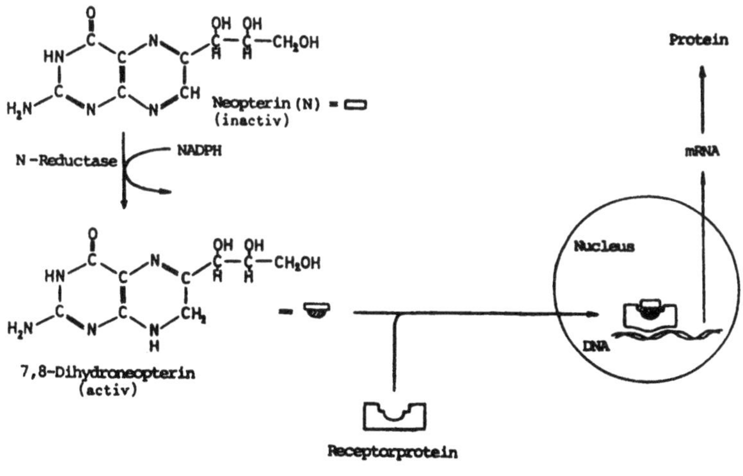

Fig.2: Mode of action of a pteridine (e.g. neopterin) in a target cell (e.g. T-lymphocyte)

dines during proliferation and differentiation, respectively, is achieved according to this hypothesis at the translation level.

A further intriguing idea is that regulation is affected in target cells. For this purpose a pteridine-binding receptor is required; one possible mode of action is shown in figure 2. The inactive pteridine is reduced, if necessary, to the active 7,8-dihydropteridine. This active pteridine combines with a specific cytoplasmic protein receptor. Then the pteridine receptor complex is transported into the cell nucleus and binds to distinct chromatin sites, where transcription is initiated. These hypotheses on the connection between pteridine release and proliferation or differentiation might stimulate further studies of these problems. This seems a very challenging question which should be answered in the next few years.

References

1. Hausen, A., Wachter, H.: J. Clin. Chem. Clin. Biochem.: in press (1982)

2. Wachter, H., Hausen, A., Grassmayr, K.: Hoppe-Seyler's Z. Physiol. Chem., 360, 1957 - 1960 (1979)

3. Fuchs, D., Hausen, A., Huber, Ch., Margreiter, R., Reibnegger, G., Spielberger, M., Wachter, H.: Hoppe-Seyler's Z. Physiol. Chem., 363, 661 - 664 (1982)

4. Huber, Ch., Fuchs, D., Hausen, A., Margreiter, R., Reibnegger, G., Wachter, H.: in preparation

5. Loidl, P., Fuchs, D., Gröbner, P., Hausen, A., Reibnegger, G., Wachter, H. In: Biochemical and Clinical Aspects of Pteridines, pp. 257 - 266, Eds.: H. Wachter, Ch. Curtius, W. Pfleiderer. Walter de Gruyter, Berlin - New York (1982)

6. Gerisch, G., Blank, G., Fuchs, D., Hausen, A., Reibnegger, G., Schweiger, M., Wachter, H. In: Biochemical and Clinical Aspects of Pteridines, pp. 253 - 256, Eds.: H. Wachter, Ch. Curtius, W. Pfleiderer. Walter de Gruyter, Berlin - New York (1982)

7. Wachter, H., Hausen, A., Reider, E., Schweiger, M.: Naturwissenschaften 67, 610 - 611 (1980)

8. Kahlson, G. and Rosengren, E. Biogenesis and Physiology of Histamine. Edward Arnold, Ltd., London (1971)

6

PTERIDINES IN METABOLIC DISEASES

ENDOGENOUS DEPRESSION AND PARKINSON'S DISEASE:

THERAPEUTIC TRIALS WITH TETRAHYDROBIOPTERIN

H.-Ch. Curtius, *H. Müldner, and A. Niederwieser

Universitäts-Kinderklinik, Steinwiesstrasse 75, CH-8032 Zürich
*Landesnervenklinik Alzey, Dautenheimer Landstr. 66
D-6508 Alzey

Summary

Three patients with endogenous depression and two patients with Parkinson's disease were treated with tetrahydrobiopterin (BH_4) in single oral doses of 0.9 - 1 g. In two of the patients with endogenous depression and in both of the Parkinson's patients, the corresponding main clinical symptoms improved or even disappeared between 4 and 10 hours after administration of the cofactor.

These preliminary observations suggest that BH_4 administration may provide an alternative method for the treatment of such diseases.

Biochemical and Clinical Aspects of Pteridines · Volume 1
Cancer · Immunology · Metabolic Diseases

Introduction

L-erythro-5,6,7,8 tetrahydrobiopterin (BH_4) is the natural cofactor of phenylalanine-4-hydroxylase (EC 1.14.16.1), t yrosine-3-hydroxylase (EC 1.14.16.2) and tryptophan-5-hydroxylase (EC 1.14.16.4); tyrosine- and tryptophan-hydroxylase are the initial and rate-limiting enzymes in the formation of the neurotransmitters dopamine and serotonin, respectively. From biochemical hypotheses of endogenous depression (1-4) and Parkinson's disease (5-8), it is possible that certain types of endogenous depression and Parkinsonism can be influenced by BH_4 administration.

Rat experiments have shown that only about 0.2 % of the peripherically administered BH_4 penetrates the blood-brain barrier (9, 10, 11). Despite this fact, we could show that in children suffering from a BH_4 deficiency caused by a defect of the enzyme dihydrobiopterin synthetase, administration of a single oral dose of BH_4 (10 - 20 mg/kg) resulted in a disappearance of the clinical symptoms for 4 days (12). Biochemical analyses demonstrated that the concentration as well as the urinary excretion rate of serotonin and dopamine were normalized. The abnormally high values of the urinary neopterin were also decreased for 2 - 3 days. A dose dependent stimulation of serotonin and dopamine production was observed. These ovservations led to the assumption that BH_4 may be effective in Morbus Parkinson and in some patients with endogenous depression.

Patients

From 3 male patients I, II, III, aged 43, 39 and 29 years respectively, patient I suffered from an agitated and patients II and III from an inhibited endogenous depression. Patients I and II had formerly been treated with different antidepressant drugs and neuroleptics without much favorable clinical response. In case II clinical improvement had been achieved 1 - 3 weeks after administration of anti-depressants. However, single doses of anti-depressant drugs were without clinical effect in all the cases.

Two patients, IV, V (72 years, female and 62 years, male) with M. Parkinson were partially compensated by treatment with L-dopa and bromocriptinmesilate (Pravidel Sandoz). Without treatment, clinically one observed hypokenesia, rigidity, tremor and a distinct quality of depressed mood.
Therapy of the patients with endogenous depression was stopped 10 days before and of those with M. Parkinson 2 days before the trials and they were on a catecholamine-low diet. None of them had a significant organic illness at the time, as evidenced by physical examination and routine laboratory tests; serum iron concentrations were normal in all cases.

Materials and Methods

BH_4.2HCl was obtained from Dr. B. Schircks, Schachenstr. 4, CH-8907 Wettswil a.A.

Biopterin was measured in urine by high-performance liquid chromatography after oxidation with MnO_2. Dopamine and serotonin were analyzed by mass fragmentography.

$BH_4 \cdot 2HCl$, 0.9 - 1 g, was administered in a glass water together with 100 mg of ascorbic acid one hour before breakfast.

Results

The most common symptoms of Parkinson's disease and endogenous depression are listed in table 1.
Patients Nr. I with agitated endogenous depression did not respond to BH_4. Patients Nr. II and III with inhibited endogenous depression showed a rapid improvement of most of the symptoms. Symptoms Nr. 1,2,3, and 5 improved in both patients significantly. Symptom Nr. 4 improved only in 1 of the patients. The improvement of the depressive symptoms occurred 4 to 5 h after administration of BH_4. The patients moods were elevated as well as their enviromental interest.
The patients became suddenly self-initiative and participated in the ward environment after developping an urge to take part in ward activities. The symptoms of depression reappeared 9 to 10 hours after the administration of BH_4.
As indicated in table 1, the two patients with Morbus Parkinson improved as well. 4 h after BH_4 administration they lost their hypokinesia, rigidity and partially the tremor. Also in these two patients, the effect lasted only for about 5 - 6 h. No side effects were observed. In table 2, the excretion of biopterin, free dopamine and serotonin in urine before and after administration of BH_4 are listed. In patient Nr. I who did not respond to BH_4, dopamine and serotonin excretion was rather high. In the two patients with inhibited endogenous depression responding to BH_4, dopamine excretion increased from low normal to high normal values for a short period of time.

Table 1: Therapeutic effect of tetrahydrobiopterin on 3 patients with endogenous depression and 2 patients with M. Parkinson.

++ Full improvement, +Light improvement, -Minimal or no improvement.

Symptoms	ENDOGENOUS DEPRESSION N=3			M.PARKINSON N=2		
	Pat. I	Pat.II	Pat.III	Pat.IV	Pat.V	
Distinct quality to depressed mood	-	++	++	+	+	(1)
Lack of reactivity to environmental changes	-	+	++	-	-	(2)
Pervasive loss of interest or pleasure	-	++	+	+	+	(3)
Feeling of self-reproach or excessive or inappropriate guilt	-	+	-			(4)
Psycho motor retardation (endogen.depression)	-	+	++			(5)
Hypokinesia (M.Parkinson)				++	++	(6)
Tremor				+	+	(7)
Rigidity				++	++	(8)

A change in the symptoms a) Mood regularly worse in the morning,
b) Early morning awakening, and
c) Poor appetite could not be judged because of the short period of duration of the effect.

Table 2: Excretion of biopterin, free dopamine and serotonin in urine before and after administration of BH_4.2HCl in 3 patients with endogenous depression.
Patients I suffered from an agitated depression and did not respond clinically.

Metabolites in urine		Patient I				Patient II					Patient III				Controls
		hours after 1 g of $BH_4 \cdot 2HCl$													
	Units	0	2	4	8	0	2	4	8-10	12	0-1	2	4-5	12	(range)
Biopterin *	mmol / mol creat.	1.51	2.51	2.09	3.59	0.36	1.36	6.71	6.04	0.91	-	-	-	-	0.31- 1.09
Dopamine	umol / mol creat.	310	360	-	210	147	208	220	220	210	78	180	200	58	70 -170
Serotonin	umol / mol creat.	100	69	84	85	30	42	60	40	34	24	34	57	61	22 - 60

* After oxidation of reduced forms by manganous dioxide at pH 1.

Discussion

Peripherically administered BH_4 enters the blood brain barrier only to a very small extent (9,10,11). On the other hand, high dose treatment (10 - 20 mg/kg BH_4) of BH_2-deficient PKU patients caused a central effect which lasted for 4 days and enhanced the serotonin and dopamine excretion in urine even above the normal range (12).
In patients with Morbus Parkinson it has recently been shown that BH_4 is decreased significantly in cerebral spinal fluid (13). We could prove (table 1) that in some patients with endogenous depression and rather low urinary concentrations of free serotonin and dopamine BH_4 administration appears to be effective for a short period of time (patient II and III). Patient Nr. I with rather high urinary excretion of serotonin and dopamine did not respond to BH_4.

It must be emphasized that the conventional methods for proving the action of neuropsychopharmaca such as rating scales could not be applied in our experiments due to the short duration of the effect. The apparent beneficial effect of BH_4 in 2 patients with endogenous depression indirectly supports the hypothesis that the availability of dopamine and serotonin in the brain may be diminished in some case.

Acknowledgement

This work was supported by the Swiss National Science Foundation project no. 3815-0.79.

References

1. Schildkraut, J.J.: Amer. J. Psychiatry 122, 509-522 (1968).

2. Usdin, E. Weiner, N., Youdim, M.B.H. (eds): Function and Regulation of Monoamine Enzymes. Basic and Clinical Aspects. Macmillan Publ. Inc., London (1981).

3. Haber, B., Gabay, S., Issidorides, M.R., Alivisatos, S.G.A. (eds): Serotonin. Current Aspects of Neurochemistry and Function. Adv. exper. Med. Biol. 133, (1981).

4. Hayaishi, O., Ishimura, Y., Kido, R. (eds): Biochemical and Medical Aspects of Tryptophan Metabolism. Developments in Biochemistry, 16, (1981).

5. Birkmayer, W., Hornykiewicz, O. (eds): Advances in Parkinsonism. Editiones Roche (1976).

6. Siegfried, J. (ed): Parkinson's Disease. Rigidity, Akinesia, Behavior, Vol. 1 and 2, Hans Huber Publ., Bern (1973).

7. Birkmayer, W., Riederer, P.: Die Parkinson-Krankheit. Biochemie, Klinik, Therapie. Springer-Verlag Wien (1980).

8. Rinne, U.K., Klingler, M., Stamm, G., (eds): Parkinsons Disease. Current Progress, Problems and Management. Elsevier (1980).

9. Gal, E.M., Hanson, G., Sherman, A.: Neurochem. Res. 1, 511-523 (1976).

10. Kapatos, G., Kaufmann, S.: Science 212, 955-956 (1981).

11. Kettler, R., Bartholini, G., Pletscher, A.: Nature 249, 476-478 (1974).

12. Niederwieser, A., Curtius, H.-Ch., Wang, M., Leupold, D.: Eur. J. Pediatr. 138, 110-112 (1982).

13. Lovenberg, W., Levine, R.A., Robinson, D.S., Ebert, M., Williams, A.C., Calne, D.B.: Science 204, 624 (1979).

SCREENING FOR TETRAHYDROBIOPTERIN DEFICIENCY

A. Niederwieser, A. Matasovic, W. Staudenmann, M. Wang and H.-Ch. Curtius

Division of Clinical Chemistry, Dept of Pediatrics, University of Zurich, Steinwiesstr. 75, CH-8032 Zurich, Switzerland

Screening in patients with persistent hyperphenylalaninemia was performed a) by measurement of urinary biopterin (B), neopterin (N) and creatinine (C) and b) by measurement of plasma phenylalanine before, 4, and 8 hours after administration of tetrahydrobiopterin dihydrochloride, 7.5 mg per kg body weight orally. Tetrahydrobiopterin (BH_4) deficiency can be detected and the variants differentiated best by plotting %B = 100B/(B + N) versus B/C. In 194 newborns with hyperphenylalaninemia, 3 patients with dihydrobiopterin deficiency were found. Further 16 patients with dihydrobiopterin deficiency, 7 patients with dihydropteridine reductase deficiency and 1 patient with postulated GTP-cyclohydrolase I deficiency were within a group of 111 older, selected patients. From 260 cases with phenylalanine-4-hydroxylase deficiency, none responded to tetrahydrobiopterin. In patients with defective cofactor biosynthesis, elevated plasma phenylalanine was normalized within 4-6 hours after tetrahydrobiopterin. From 7 patients with dihydropteridine reductase deficiency, 5 responded partially to tetrahydrobiopterin and 2 did not respond.

Introduction

Screening is a diagnostic activity performed to discover a disorder which otherwise would not be recognized easily in an unselected sample of the population. Preconditions for a screening program are that the disease is severe, without specific symptoms and treatable, and that a suitable method is available which can be applied before irreversible harm to the patient has occurred (1). Neonatal screening for inborn errors of metab-

Biochemical and Clinical Aspects of Pteridines · Volume 1
Cancer · Immunology · Metabolic Diseases

olism was begun around 1960 and mass screening in the population for phenylketonuria (PKU) became possible by the introduction of the bacterial inhibition test by Guthrie and Susi (2). Screening for inborn tetrahydrobiopterin (BH_4) deficiency fulfills the above requirements in so far as the disease is a severe, progressive and lethal neurological disorder (3), treatable by precursors of neurotransmitters (4) and /or cofactor substitution (5,6,7). However, it is a selected screening because such patients show a mild or severe hyperphenylalaninemia which in each case can be recognized by the mass screening program for PKU. Until recently, such atypical patients were detected only rather late because of their neurological symptoms despite treatment with low phenylalanine diet, at an age when irreversible brain damage might have already occurred.

Danks (3,7) was the first to propose a diagnostic test by administration of BH_4 and measurement of plasma phenylalanine (Phe), but at that time BH_4 was not yet available in sufficient amounts and purity. Since the discovery of our first patient with BH_4 deficiency in 1978 (8), we tried to develop reliable techniques for early detection of BH_4 deficiency in newborns with hyperphenylalaninemia. Pure BH_4 had been made available by Schircks and Viscontini (9) and was employed for diagnostic and therapeutic purposes (5,10).

Studying variants of hyperphenylalaninemia, it was ruled out that the pattern of urinary pterins was very characteristic for these disorders. The analyses of pterins after oxidation enabled us not only to detect BH_4 deficiency but also to differentiate the BH_4-deficient variants, namely, dihydropteridine reductase (DHPR) deficiency and dihydrobiopterin (BH_2) deficiency. In the beginning, we analyzed pterins only semiquantitatively using high voltage electrophoresis and paper chromatography (5,10). Trials for quantitation by gas chromatography-mass fragmentography (11) were abandoned in favour of high-performance liquid chromatography (HPLC) (12). The latter is at present the method of choice, being fast, automatic and needing no pretreatment of the sample except oxidation.

In addition to the analysis of urinary pterins, the BH_4 test is still in use because in most cases it gives an immediate answer to the physician concerned: It needs only measurement of plasma/serum Phe which can be performed everywhere. In the following, we describe the screening as it is presently performed and the results obtained.

Materials and Methods

BH_4 tablets

After many trials kindly performed by pharmacist H. Hotz (Küsnacht ZH, Switzerland) in collaboration with us, a mixture of cofactor and stabilizing agents was found which gave relatively stable tablets. Each tablet contains:

10 mg BH_4 dihydrochloride (Dr. B. Schircks, Wettswil a.A., Switzerland);
10 mg ascorbic acid (E. Merck, Darmstadt, FRG);
5 mg *N*-acetyl cysteine (Inpharzam SA, CH-6814 Cadempino, Switzerland).

The tablets have been approved by the Bundesgesundheitsamt (Ministry of Health, Berlin, FRG) for use in screening and treatment of BH_4 deficiency. They are now available from Dr. B. Schircks, distributed by Milupa AG (D-6382 Friedrichsdorf, FRG).

Diagnostic test for BH_4 deficiency

The test should be applied to all newborns or patients with persistent hyperphenylalaninemia (plasma Phe > 150 µmol/l). Before testing, the patient's plasma Phe concentration must be *constant and elevated* (>400 µmol/l, if possible), by adaptation of dietary treatment, if necessary. Newborns should be tested before institution of PKU diet. The test consists of:

A : measurement of urinary pterins and

B : measurement of plasma Phe before and after a single oral dose of BH_4 dihydrochloride, approx. 7.5 mg/kg body weight, given at elevated plasma Phe level.

BH_4 is administered 30-60 min before the next meal to ensure quick absorption. Only immediately before use, the appropriate number of BH_4 tablets are dissolved in water (e.g. for newborns 3-4 tablets à 10 mg $BH_4 \cdot 2HCl$ dissolved in 10-30 ml water). After 2 min, the suspension is administered to the patient; a stomach tube can be used. These manipulations should be done in dim light. *Blood samples* are taken for measurement of phenylalanine and tyrosine, immediately before loading (0 h) and 4 and 8 h afterwards. *Urine* is collected for measurement of pterins and creatinine, 12-0 h before and 4-8 h after loading with BH_4. The collection 4-8 h after BH_4 is used only to check correct absorption of BH_4. Protect samples from light! Aliquots of 5-10 ml are frozen immediately in plastic tubes and should be thawed only just before analysis. Samples should be shipped frozen, packed in dry ice, and by air freight, if necessary.

Measurement of metabolites

Plasma Phe was measured by automatic amino acid chromatography. Urinary pterins were measured by HPLC after oxidation with manganese dioxide (12) (see also our chapter on automatic HPLC of pterins in this book). Urinary creatinine was determined colorimetrically using the Jaffé reaction.

Results and Discussion

Because complete urine collections could not be guaranteed, total biopterin and neopterin concentrations were expressed in mmol/l creatinine. The ratio of biopterin and neopterin was expressed as % biopterin (B):

$$\%B = \frac{\text{biopterin} \cdot 100}{\text{biopterin} + \text{neopterin}}$$

Healthy newborns excreted relatively more neopterin than children and adults (12). In phenylalanine-4-hydroxylase (PH) deficient patients, the decrease in the ratio of neopterin/creatinine with increasing age was even more pronounced. These observations may indicate some correlation with growth or a progressive maturation of dihydrobiopterin biosynthesis in children. Newborn patients with PH deficiency at elevated plasma Phe excreted more pterins than healthy controls. In the individual patient, a positive correlation between biopterin or neopterin excretion and plasma Phe could be observed. For instance, dietary treatment corrected the plasma Phe of a newborn PKU patient from 1940 to 120 µmol/l and lowered biopterin from 4.7 to 1.6 and neopterin from 17 to 3.8 mmol/mol creatinine, respectively. Obviously, there is a feedback mechanism operating here.

Patients with DHPR deficiency showed the highest biopterin excretion and %B was always higher than 80 at elevated plasma Phe. Neopterin excretion was comparable to that of PH-deficient patients. This fact indicated that the metabolic block at DHPR did not provoke accumulation of a neopterin derivative, most probably because an irreversible step is involved in the metabolism between dihydroneopterin triphosphate and dihydrobiopterin.

Patients with BH_2 deficiency excreted very low amounts of biopterin and very high amounts of neopterin. All specimens analyzed gave values below 5%B.

A patient with postulated GTP-cyclohydrolase I deficiency excreted only very low amounts of pterins; %B was normal.

Typical HPLC chromatograms of urinary pterins in newborns and older children are shown in Figs. 1 and 2, respectively.

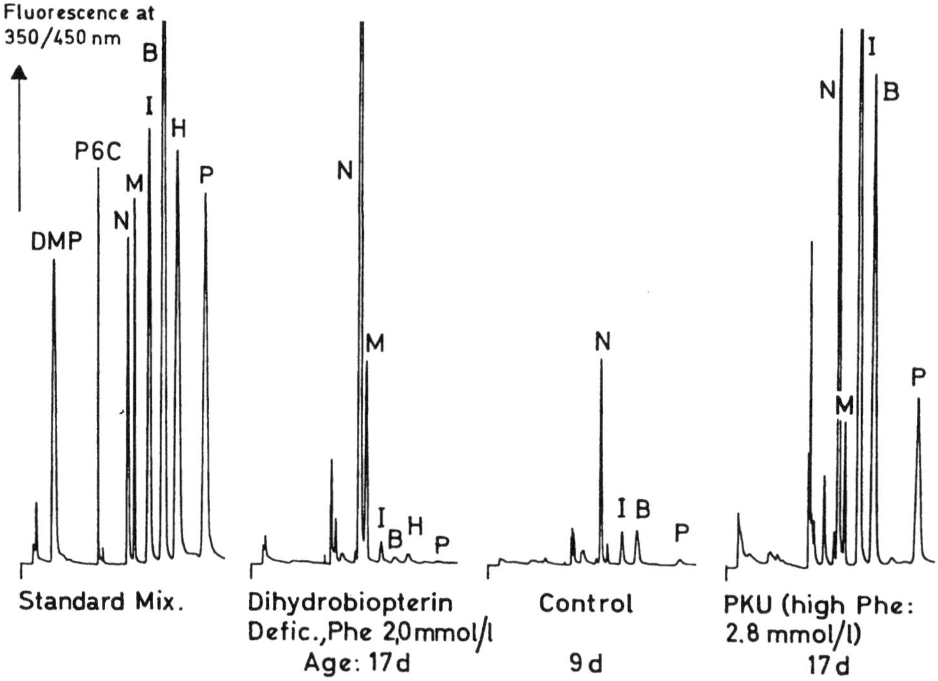

Fig. 1. High-performance liquid chromatogram of oxidized urinary pterins of newborn patients with dihydrobiopterin deficiency, phenylalanine-4-hydroxylase deficiency (PKU) as well as healthy control. The standard mixture consisted of 6,7-dimethylpterin (DMP), pterin-6-carbonic acid (P6C), neopterin (N), monapterin (M), isoxanthopterin (I), biopterin (B), 6-hydroxymethylpterin (H) and pterin (P).

To make chromatograms comparable, sample volumes equivalent to the same amount of creatinine were injected for HPLC. In BH_2 deficiency, the high concentrations of neopterin and monapterin and the very low concentration of biopterin are impressive (Fig1).

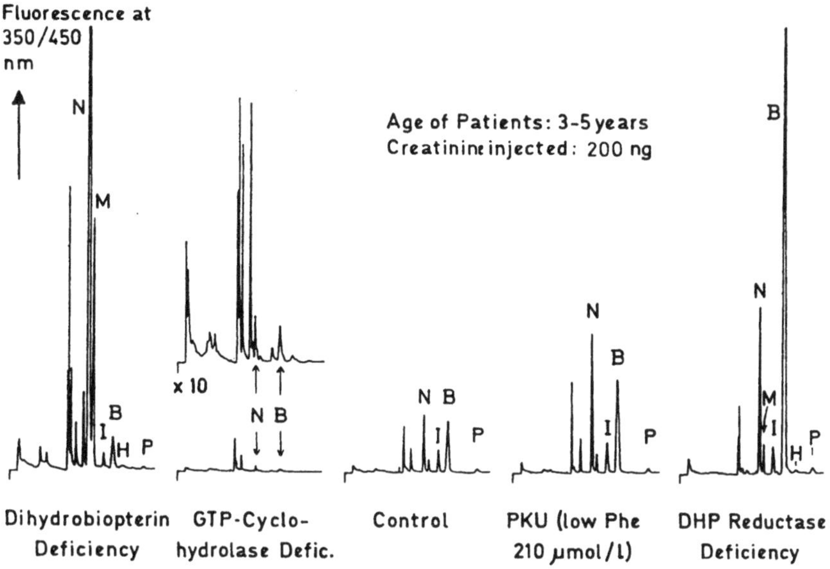

Fig. 2. HPLC of oxidized urinary pterin of older patients with deficiencies in dihydrobiopterin, GTP-cyclohydrolase, phenylalanine-4-hydroxylase (PKU) and dihydropteridine reductase (DHPR), as well as that of a healthy child. Abbreviations see figure 1.

In older patients, however, a significant concentration of biopterin was found (Fig. 2).

The origin of biopterin in such patients is still unknown: It could be of dietary origin, from intestinal flora or be formed due to a residual activity of the mutated enzyme. Traces of pterins were also detected in the patient with postulated GTP-cyclohydrolase deficiency. It is noteworthy that the chromatographic pattern of neopterin, monapterin, isoxanthopterin, biopterin and pterin became comparable to the control when a 10 times greater sample volume was analyzed (Fig. 2). The great variability of excretion of pterins in PKU depending on plasma Phe concentration is also shown in Figs. 1 and 2. At low plasma Phe concentration, PKU patients either differed slightly or not at all from healthy controls (Fig. 2). Similarly, the high biopterin excretion in DHPR deficiency (Fig. 2) decreased under a low Phe diet so far that their pattern of urinary pterins could not be distinguished anymore from that of PKU patients (not shown). Therefore, it is necessary to perform screening tests for BH_4 deficiency at elevated plasma Phe and not under low Phe diet.

The variants with BH_4 deficiency could be distinguished best when %B was plotted versus biopterin/creatinine (Fig. 3).

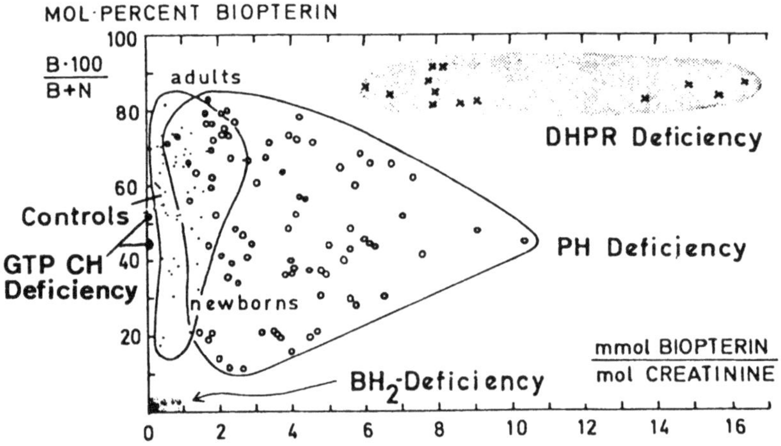

Fig. 3. Detection and differentiation of tetrahydrobiopterin-deficient variants of hyperphenylalaninemia including GTP-cyclohydrolase I (GTP CH) deficiency by urine biopterin (B), neopterin (N), and creatinine. Urine samples were collected for 12 h at elevated plasma phenylalanine concentrations, which values differed widely between patients (6). In phenylketonuric patients (phenylalanine-4-hydroxylase deficiency) a decrease in plasma phenylalanine by dietary treatment brought the values from the right to the left.

The values found in PH, DHPR and BH_2 deficiency fall into areas completely separate from each other. The values of the new variant with postulated GTP-cyclohydrolase deficiency were around 50%B, near the ordinate and also separate from other variants and controls. The abscissa (biopterin/creatinine) should be drawn in logarithmic scale to demonstrate that more clearly.

It might be expected that patients will be found, repeatedly showing values between PH and DHPR deficiency or PH and BH_2 deficiency. In the former case, measurement of DHPR in liver biopsy or skin fibroblast culture will be necessary to clarify the situation (partial deficiency in DHPR?). On the other hand, the enzymes in BH_2 biosynthesis cannot yet be measured in biopsies. However, even a partial defect in BH_2 biosynthesis will be detected by the BH_4 administration test.

The BH_4 test allowed to detect all patients with a defect in biosynthesis of biopterin including GTP-cyclohydrolase deficiency, and most of the patients with DHPR deficiency. A dramatic decrease of elevated plasma Phe was observed in patients with defective biopterin biosynthesis, and plasma Phe became normal 4-6 hours after BH_4 administration (Fig. 4). Simultaneously, plasma tyrosine rose sharply, but was normalized fast, presumably by induction of the tyrosine degrading enzymes. One patient (M.A.) with DHPR deficiency responded to BH_4 very well (5), 4 others responded partially, by a slight decrease in plasma Phe concentration e.g. from 6 to 4 mg/dl, and 2 patients (out of 7 investigated) did not respond at all. This observation cannot be explained yet.

In patients with defective biopterin biosynthesis, catalytically small amounts of cofactor are sufficient for normalization, because the regeneration cycle is intact. In DHPR deficiency, more BH_4 is needed although in most patients it may be less than the theoretical stoichiometric amounts. The observation that hyperphenylalaninemia is only moderate in DHPR-deficient

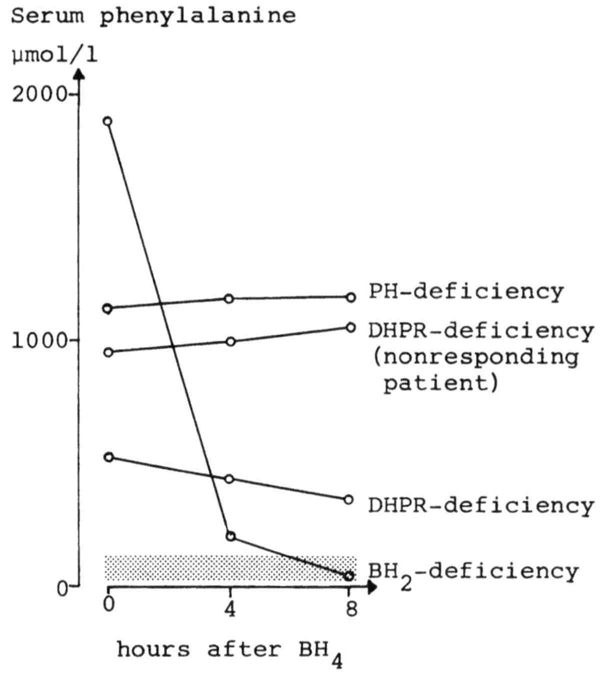

Fig. 4. Typical results of tetrahydrobiopterin (BH_4) test in patients with deficiency in phenylalanine-4-hydroxylase (PH), dihydropteridine reductase (DHPR) and dihydrobiopterin (BH_2). Two of 7 patients with DHPR deficiency did not respond to administration of $BH_4 \cdot 2HCl$, 7.5 mg/kg, the others responded partially. The hatched area corresponds to the normal range.

patients (3,5) indicates that these patients are able to reduce by other means in the periphery part of the quinonoid BH_2 (or the 7,8-dihydrobiopterin derived therefrom) which has been formed during the hydroxylation of aromatic amino acids. It is known that thiols like reduced glutathione are able to reduce quinonoid BH_2 nonenzymatically and BH_2 is accepted slowly (at higher K_M) by dihydrofolate reductase.

Our results of screening for BH_4 deficiency are compiled in Table 1.

TABLE 1

Results of screening children with hyperphenylalaninemia for tetrahydrobiopterin (BH_4) deficiency, using BH_4 administration test and analysis of urinary pterins by HPLC.

Children investigated		found	
194	newborns with hyperphenylalaninemia	3	BH_2-deficient
111	older, selected patients	16	BH_2-deficient
		7	DHPR-deficient
		1	GTP-cyclohydrolase-deficient
260	phenylalanine-4-hydroxylase-deficient		no one responding to BH_4, 7.5 mg/kg

It is noteworthy that none of the 260 PH-deficient patients investigated responded to BH_4. That means, in none of these patients, mutation of PH effected only weakening of cofactor binding. Such K_M mutants are well known, however, in other co-factor-dependent enzyme systems, as demonstrated in vitamin dependencies e.g. of methyl malonic aciduria, cystathioninuria, primary hyperoxalosis, etc. (13). The BH_4 test would enable us to detect such K_M mutant PKU patients in future, if they exist at all. Table 1 is the begin of statistics on the incidence of the rare variants suffering from BH_4 deficiency. In 1977, the incidence of BH_4 deficiency was estimated to be 1-3% in PKU patients (3). Our findings of 3 cases in 194 newborns with hyperphenylalaninemia is within this estimate. It is already clear that the incidence of BH_2 deficiency is higher (by a factor of approx. 2) than DHPR deficiency.

We might expect still other variants of defects in biosynthesis of BH_4 and are hoping to detect them by our screening procedure. It must be emphasized that this screening procedure is non-invasive, needs no liver biopsy but only a small urine specimen for measurements of pterins and blood samples (capillary blood would be sufficient) for measurement of phenylalanine and tyrosine.

Two of the BH_2-deficient patients had normal or only slightly elevated serum phenylalanine concentrations in the newborn period, possibly because of rather low protein and sufficient BH_4 intake during breast-feeding. Consequently, the cut off phenylalanine level for selected screening of BH_4 deficiency should be set as low as possible.

Every patient with abnormal excretion of pterins in urine should be investigated further for the presence of unusual pterins, like 3'-hydroxysepiapterin (14). It might be expected that an intermediate in biopterin biosynthesis will be detected which could help to clarify the mechanism of biosynthetic pathway.

Acknowledgements

The data presented was made possible through valuable collaboration with many physicians in various pediatric clinics both throughout Europe and abroad. We are very grateful to pharmacist H. Hotz (Küsnacht ZH, Switzerland) who, at great personal time and expense, was able to successfully formulate tablets of BH_4 dihydrochloride for these studies. We thank Mr. E. Wetzel and Mr. W. Scherrer for skillful technical assistance. This work was supported by the Swiss National Foundation, project no. 3.815-0.79.

References

1. Bickel, H., Guthrie, R., Hammersen, G.(eds.): "Neonatal Screening for Inborn Errors of Metabolism". Springer-Verlag Berlin (1980).

2. Guthrie, R., Susi, A.: A simple phenylalanine method for the detection of phenylketonuria in large populations of newborn infants. Pediatrics 32, 338 (1963).

3. Danks, D.M., Bartholomé, K., Clayton, B.E., Curtius, H., Gröbe, H., Kaufman, S., Leeming, R., Pfleiderer, W., Rembold, H., Rey, F.: Malignant hyperphenylalaninaemia-current status (June 1977). J. Inher. Metab. Dis. 1, 49-53 (1978).

4. Bartholomé, K., Byrd, D.J.: L-Dopa and 5-hydroxytryptophan therapy in phenylketonuria with normal phenylalanine-hydroxylase activity. Lancet II, 1042 (1975).

5. Curtius, H.-Ch., Niederwieser, A., Viscontini, M., Otten, A., Schaub, J., Scheibenreiter, S., Schmidt, H.: Atypical phenylketonuria due to tetrahydrobiopterin deficiency. Diagnosis and treatment with tetrahydrobiopterin, dihydrobiopterin and sepiapterin. Clin. Chim. Acta 93, 251-262 (1979).

6. Niederwieser, A., Curtius, H.-Ch., Wang, M., Leupold, D.: Atypical phenylketonuria with defective biopterin metabolism. Monotherapy with tetrahydrobiopterin or sepiapterin, screening and study of biosynthesis in man. Eur. J. Pediat. 138, 110-112 (1982).

7. Danks, D.M., Cotton, R.G., Schlesinger, P.: Tetrahydrobiopterin treatment of variant forms of phenylketonuria. Lancet II, 1043 (1975).

8. Schaub, J., Däumling, S., Curtius, H.-Ch., Niederwieser, A., Bartholomé, K., Viscontini, M., Schircks, B., Bieri, J.H.: Tetrahydrobiopterin therapy of atypical phenylketonuria due to defective dihydrobiopterin biosynthesis. Arch. Dis. Childh. 53, 674-676 (1978).

9. Schircks, B., Bieri, J.H., Viscontini, M.: Eine neue, regiospezifische Synthese von L-Biopterin. Helv. Chim. Acta 60, 211 (1977).

10. Niederwieser, A., Curtius, H.-Ch., Bettoni, O., Bieri, J., Schircks, B., Viscontini, M., Schaub, J.: Atypical phenylketonuria caused by 7,8-dihydrobiopterin synthetase deficiency. Lancet I, 131-133 (1979).

11. Niederwieser, A., Bieri, J., Blehova, B., Curtius, H.-Ch., Matasović, A., Rey, F., Schircks, B., Viscontini, M.: Urinary pterins in atypical phenylketonuria (Abstract). Pediatr. Res. 14, 175 (1980).

12. Niederwieser, A., Curtius, H.-Ch., Gitzelmann, R., Otten, A., Baerlocher, K., Blehova, B., Gröbe, H., Rey, F., Schaub, J., Scheibenreiter, S., Schmidt, H., Viscontini, M.: Excretion of pterins in phenylketonuria and phenylketonuria variants. Helv. Pediat. Acta 35, 335-342 (1980).

13. Rosenberg, L.E.: Vitamin-responsive inherited metabolic disorders. Advances in Human Genet. 6, 1-74 (1976).

14. Niederwieser, A., Matasović, A., Curtius, H.-Ch., Endres, W., Schaub, J.: 3'-Hydroxysepiapterin in patients with dihydrobiopterin deficiency. FEBS Lett. 118, 299-302 (1980).

TETRAHYDROBIOPTERIN MONOTHERAPY IN TWO SIBLINGS WITH DIHYDROBIOPTERIN DEFICIENCY

D. Leupold

Department of Pediatrics, University of Ulm
D-79oo Ulm/Do., Germany

M. Wang and A. Niederwieser

Department of Pediatrics, University of Zürich
CH-8o32 Zürich, Switzerland

Introduction

In classical phenylketonuria (PKU) due to phenylalanine hydroxylase deficiency mental retardation is prevented by early restriction of phenylalanine in the diet. However, about 1 - 3 per cent of PKU patients show progressive neurological deterioration despite adequate therapy (1). During the last 6 years this formerly termed "malignant hyperphenylalaninemia" has been proven to originate from cofactor-, i. e. tetrahydrobiopterin (BH_4)-deficiency.
The metabolic error may be either defective dihydrobiopterin (BH_2) synthesis (2, 3, 4, 5) or defective regeneration due to a deficient activity of dihydropteridine reductase (DHPR) (1, 3, 6).

Since BH_4 is cofactor not only of phenylalanine hydroxylase (7) but also of tyrosine-3-hydroxylase (8) and tryptophan-5-hydroxylase (9) BH_4 deficiency causes severe cerebral damage in consequence of impaired synthesis of the neurotransmitters dopamine and serotonin.

In contrast to classical PKU clinical symptoms begin during the first weeks of life and consist in feeding difficulties,

Biochemical and Clinical Aspects of Pteridines · Volume 1
Cancer · Immunology · Metabolic Diseases

hypersalivation and muscular dystonia with moderate to marked truncal hypo- and hypertonia and rigidity of the limbs. Later on a greasy skin and, occasionally, disturbed regulation of body temperature - hypo- as well as hyperthermia - are observed (2, 3).
The infants have no social contact, show progressive neurological deterioration and - without therapy - usually die during the first years of life.

The first successful treatment, consisting in substitution of the immediate neutrotransmitter precursors L-dopa and 5-hydroxytryptophan (5-HTP) in combination with dietary phenylalanine restriction was reported by Bartholomé and Byrd (1o) in a patient with defective BH_2 biosynthesis.

Low doses of BH_4 very effectively decrease the serum phenylalanine (phe) concentration (11) and make dietary restriction unnecessary. As early as 1975 Smith et al. (12) suggested BH_4 administration as a therapeutic principle. It has been, however, doubted that BH_4 would pass the blood-brain barrier (13). Therefore, the usual recommended therapy for patients with BH_2 deficiency is a combination of BH_4, the neurotransmitter precursors L-dopa plus 5-HTP plus an aromatic amino acid decarboxylase inhibitor, (Carbidopa). The dosage varies and is regulated according to clinical response (Table 1).
Gal et al. (14) reported that peripherally administered BH_4 penetrated into the brain in very small amounts. Very recently this was confirmed by Kapatos and Kaufman (15). They suggested that treatment with high doses of BH_4 would probably increase the intracellular BH_4 concentration sufficiently to correct the deficiency. Following this suggestion we treated two patients with BH_4 monotherapy and report the results.

Case reports

Case 1, Y. Z., born september 1976, was published earlier (16)

Table I

Atypical PKU, Defect in Dihydrobiopterin Biosynthesis

Conventional Therapy	mg / kg b.w. / day
BH_4	ca. 2,5
L-DOPA	1o - 15
5-Hydroxytryptophan	ca. 4
Decarboxylase Inhibitor	1 - 2

Treatment with BH_4, 3o mg per day, and neurotransmitter substitution (L-dopa 75 mg, 5-HTL 75 mg, carbidopa 12 mg) in three divided doses was given very irregularly and the girl frequently needed hospitalization. About two days after withdrawal of medication she got into a crisis, which began with ptosis, frequent yawning, loss of mimic and pronounced muscular hypotonia, followed by ataxia and increasing sleepiness. One of these crises developed, although neurotransmitter precursors had been given correctly and only BH_4 was missing.
Monotherapy was started during a crisis with 4oo mg BH_4 (22 mg/kg bw) given by mouth. After 3 - 4 hours the patient responded dramatically: all symptoms vanished, the girl was fully active and remained so for four days. The same effect was obtained with 11 mg/kg BH_4 and L-sepiapterin in a dosage as low as 2.75 mg/kg b.w.. We decided to continue the therapy with a daily single dose of 5o mg BH_4 (2.5 mg/kg b.w.).
The patient is now on monotherapy for 12 month. She shows a slight muscular hypotonia but is otherwise normal. Only during a viral infection with pneumonitis and gastroenteritis neurological symptoms reappeared. Whether this resulted from impaired absorption or increased requirement is not clear. However, a fourfold dosage of BH_4 was needed for restoration. In order to prevent any risk for the patient, the daily BH_4 dosage was increased to 5.5 mg/kg b.w..

Case 2, Y. Ö., born november 1981, is the brother of Y, Z.. Pregnancy and delivery were uneventful. During the first 17 days of life he was breast-fed. At the age of 18 days of life he was admitted to our hospital, because his plasma phe at the age of 6 days was elevated (o.848 mmoles/l).
On admission he appeared completely healthy without any neurological abnormalities. Plasma phe was excessively high (2.o15 mmoles/l) but other routine laboratory data were normal.
BH_4 therapy was started with lo mg/kg b.w..
At the age of 6 weeks short dystonic episodes with opisthotonic

head posturing, limb rigidity and cloni of the feet were observed. These symptoms disappeared after institution of neurotransmitter therapy in a final dose of L-dopa 2.o mg/kg b.w., Carbidopa o.2 mg/kg b.w. and 5-HTP 1.5 mg/kg b.w.. After 5 weeks this regimen was discontinued without reappearance of neurological symptoms. The boy is now 7 month old and normal.

Biochemical investigations and results

Urine samples were investigated for pterin metabolites (17) dopamine (18) and serotonin (19).

Case 1, Y. Z.
The effect of BH_4 was tested with 4oo mg per day (22 mg/kg b.w.), 2oo mg per day (11 mg/kg b.w.) and with L-sepiapterin, 5o mg per day (2.75 mg/kg b.w.).
Results are shown in table 2 and fig. 1.
With either 4oo or 2oo mg of BH_4 biopterin excretion increased to about fivefold normal levels within 4 hours after administration neopterin decreased to about one third initial value. Serotonin excretion was elevated above normal levels with 4oo mg, but reached only normal range with 2oo mg BH_4.Dopamine, although estimated sporadically, seemed to be in the normal range. This effect lasted for two days, than neopterin rose again, in parallel with plasma phe, and biopterin and serotonin decreased below the range of controls. 5o mg L-sepiapterin had the same lowering effect on neopterin but biopterin increase was less and was subnormal 24 Hours after application. Serotonin was in the normal range for two days. Long-term treatment with BH_4, 5o mg per day in one single dose kept plasma phe below o.1 mmole/l and biopterin in the range of controls, and neopterin below 5 µmoles/mol creatinine.

Case 2, Y. Ö.
In fig. 2 the effect of BH_4, lo mg/kg b.w. on plasma phe is

TABLE II

Case 1, Y.Z., Concentration of urine biopterin (B), Neopterin (Ne), serotonin, dopamine and plasma phenylalanine (phe) after high-dose BH_4.

BH_4	B	Ne	serotonin	dopamine	phe
	(mmoles/mol	creatinine)	(μmoles/mol	creatinine)	(mmoles/l)
pre	o.11	13.5	41		o.4oo
post 22 mg/kg bw					
o-4 h	1.14	13.6o	86		o h o.697
4-8 h	11.3o	11.3o	47o	44o	
8-12 h	5.9o	8.9o	16o		
12-24 h	1.o7	4.16	11o		24 h o.o36
48 h	o.43	5.66	11o		
72 h	o.13	8.65	59		
Controls (12 month - 5 years)	o.53-1.67	o.23-1.71	7o-16o	36o-94o	o.o3-o.o7

Figure 1 : Case 1, Y.Z., concentration of plasma phenylalanine (Phe) and urinary excretion of neopterin (Ne), biopterin (B) and seretonin (Ser) during therapy with tetrahydrobiopterin (BH_4) and L-sepiapterin.

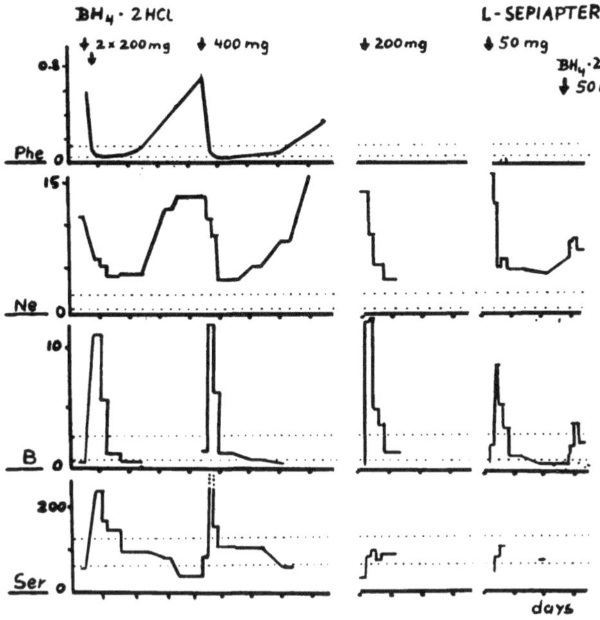

Figure 2 : Concentration of plasma phenylalanine and tyrosine in case 2 (Y.Ö.)after 3o mg BH_4 p.o., compared with two patients with classical PKU.

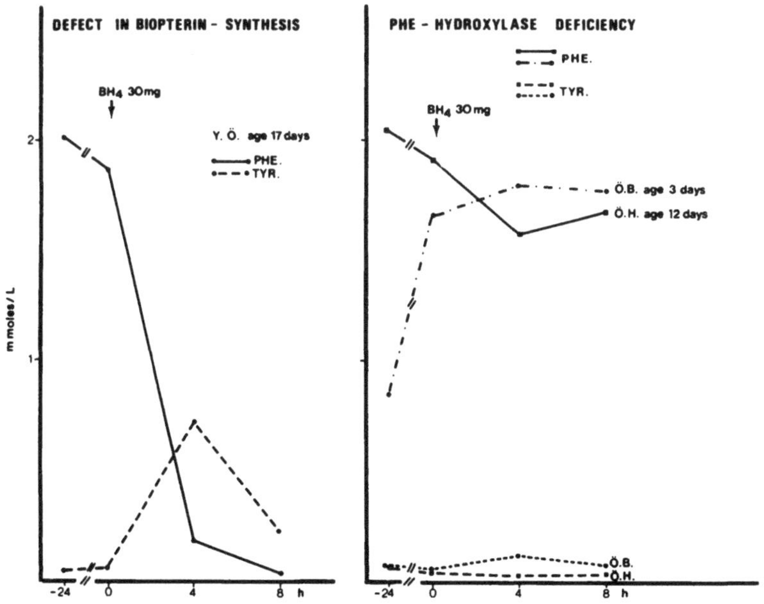

TABLE III

Case 2, Y.Ö., Concentration of urine biopterin (B), neopterin (Ne), serotonin, dopamine and plasma phenylalanine (phe) during BH_4 monotherapy.

age (days)	BH_4 therapy (mg/kg bw)	B (mmoles/mol creatinine)	Ne (mmoles/mol creatinine)	serotonin (µmoles/mol creatinine)	dopamine (µmoles/mol creatinine)	phe (mmoles/l)
6	ø	o.18	14.7	n.d.	n.d.	o.848
16	ø	<o.o7	32.8	<35	245	2.o15
18	lo					
o-2 h		4.o	28.4	<45	38o	o h 1.873
2-4 h		13.2	3o.2	65	395	
4-6 h		16.5	28.5	n.d.	n.d.	4 h o.188
6-8 h		13.o	22.9	175	13oo	
8-24 h		4.38	13.3	loo	looo	8 h o.o31
21	lo	8.78	5.96	85	77o	o.o32
28	lo	9.48	4.6o	n.d.	n.d.	n.d.
38	lo	3.72	3.87	loo	49o	o.o79
53	lo	3.76	4.43	13o	n.d.	o.o42

x n.d. = not done

controls 14-9o days		o.45-1.17	1.o-2.88	9o-65o	75-19o	o.o3-o.o7

shown in this patient with BH_2 deficiency in comparison to two infants with classical PKU due to phe hydroxylase deficiency. Plasma phe dropped dramatically within 4 hours and was normal 8 hours after BH_4 application. The excessive hydroxylation of phe led to marked hypertyrosinemia.
Table 3 gives the urinary excretion of pterine metabolites, dopamine and serotonin.
Biopterin increase in this infant is similar to that found in his older sister. Neopterin which was much higher in the infant, declined within 24 hours to about 3o per cent of the initial value, but took longer before it reached the same low levels which were obtained in the sister. Dopamine and serotonin excretion normalized completely and remained in normal range, even when neurological symptomes were observed.

Discussion

In inborn errors of biopterin metabolism severe cerebral damage results from defective neurotransmitter biosynthesis.The hitherto recommended therapy therefore besides BH_4 consists in replacement of the neurotransmitter precursors L-dopa and 5-HTP and an aromatic aminoacid decarboxylase inhibitor (like Carbidopa).
Side-effects from L-dopa are well known in patients treated for Parkinson's disease. They include dyskinesia, hallucinations and psychoses (2o, 21). Some of the dyskinesias resemble the original symptoms of BH_2 deficiency. Side-effects from 5-HTP and Carbidopa were reported recently (22) in two patients who developed a scleroderma-like illness during long-term treatment for intention myoclonus.

In patients with defective BH_2 synthesis there actually is a lack of intracellular BH_4. High-dose BH_4 treatment, therefore, would be a causal therapy.
Monitoring of intracerebral metabolic processes in the pa-

tients is difficult. The complete and prompt clinical recovery of patient Y. Z., 3 - 4 hours after BH_4 administration strongly suggests endogenous neurotransmitter production. The clinical picture is in accordance with the increase of dopamine and serotonin excretion in the urine. We are quite aware, that urinary excretion of neurotransmitters is a very peripheral parameter of brain metabolism. However, this method of control has to be used, since frequent investigation of cerebrospinal fluid, which would reflect the intracerebral situation much better, is not practicable.
From our experience urinary excretion of neopterin serves as a sensible parameter in long-term follow-up, because with insufficient dosage of BH_4 an immediate increase of neopterin is demonstrable.

Our successful treatment with high-dose BH_4 in two children suffering from BH_2 deficiency seems to justify a trial in further patients.

References

1. Danks, D.M., Bartholomé, K. et al.: J. Inher. Metab. Dis. 1, 49-53 (1978).
2. Bartholomé, K., Byrd, D.J. et al.: Pediatrics 59, 757-761 (1977).
3. Rey, F., Harpey, J.P. et al.: Arch. franç. Pediat. 34, lo9-12o (1977).
4. Kaufman, S., Berlow, S. et al.: New Engl. J. Med.299, 673-679 (1978).
5. Niederwieser, A., Curtius, H.Ch. et al.: Lancet I, 131-133 (1979).
6. Kaufman, S., Holtzman, N.A. et al.: New Engl. J. Med. 293, 785-79o (1975).
7. Kaufman, S.: J. Biol. Chem. 23o, 931 (1958).
8. Lloyd, T., Weiner, N.: Molec. Pharmacol. 7, 569 (1971).
9. Hosoda, S., Glick, D.: J. Biol. Chem. 241, 192 (1966).
lo. Bartholomé, K., Byrd, D.J.: Lancet II, lo42-lo43 (1975).

11. Danks, D.M., Cotton, R.G.H., Schlesinger, P.: Lancet II, 1o43 (1975).

12. Smith, L., Clayton, B.E., Wolff, O.H.: Lancet I, 11o8-1111 (1975).

13. Kettler, R., Bartholini, G., Pletscher, A.: Nature 249, 476-478 (1974).

14. Gal, E.M., Hanson, G., Sherman, A.: Neurochem. Res. 1, 511-523 (1976).

15. Kapatos, G., Kaufman, S.: Science 212, 955-956 (1981).

16. Schaub, J., Däumling, S. et al.: Arch. Dis. Childh. 53, 674-675 (1978).

17. Niederwieser, A., Curtius, H.-Ch. et al.: Helv. paediat. Acta 35, 335-342 (198o).

18. Curtius, H.-Ch., Wolfensberger, M. et al.: J. Chromatogr. 99, 529-54o (1974).

19. Curtius, H.-Ch., Farner, H., Rey, F.: J. Chromatogr. 199, 171-179 (198o).

2o. Hunter, K.R., Shaw, K.M. et al.: Lancet II, 929-931 (1973).

21. Shaw, K.M., Lees, A.J., Stern, G.M.: Q.J.Med. 49, 283-293 (198o).

22. Sternberg, E.M., Van Woert, M.H. et al.: N. Engl. J. Med. 3o3, 782-787 (198o).

Keye-words: Atypical phenylketonuria, BH_2 deficiency, therapy, tetrahydrobiopterin.

RELATIONSHIPS BETWEEN PHENYLALANINE AND BIOPTERIN METABOLISMS.

J.L.DHONDT

Laboratoire de Biochimie, Faculté libre de Médecine,56 rue du Port, 59046 Lille Cédex (France)

J.P.FARRIAUX

Service de Génétique et Maladies héréditaires du métabolisme, Cité Hospitalière,59037 Lille Cédex (France)

The enzymatic hydroxylation of Phenylalanine requires tetrahydrobiopterin (THB) as a cofactor. New variants of hyperphenylalaninemia have been described to be caused by a defective synthesis of THB (1,2). In addition, abnormal concentrations of pteridine have also been shown in serum and urine from PKU patients (Phenylalanine-hydroxylase deficiency)(2-4). However, the mechanism of these variations was not yet completely elucidated.

The aim of this work was to precise the relationships between phenylalaninemia and pteridine metabolism in man and in animal models.

Material and Method:

Patients: Sera and urines were obtained from 25 PKU patients and from 50 controls. An oral phenylalanine load (100mg/kg or 200mg/kg) was given to 32 healthy subjects (7 days-30 years), sera were collected at time 0 and 2 hours and urine samples before the load and for the following 4 hours.

Animal models: Adult male rats were injected intraperitoneally with 300 mg/kg of phenylalanine and sacrificed 30 minutes later. A control group was injected with an isotonic saline solution.

Biochemical and Clinical Aspects of Pteridines · Volume 1
Cancer · Immunology · Metabolic Diseases

Isolated perfused canine livers (650-700 g) were obtained according to the method of Kestens (5). When the normal haemodynamic and metabolic status was reached, a phenylalanine load (1.6 and 6.0 g) was performed.

Method: Pteridines concentrations in serum and liver were measured according to the method previously described (6).

Results :

In regard to the developmental changes of pteridine concentration in biological fluids during childhood (4), age-paired comparisons between controls and PKU patients were achieved.

Plotting pteridine elimination versus phenylalaninemia in two selected age groups : 10 days- 1 month and 6 - 10 years (fig.1), it appeared that biopterin elimination increased with the phenylalaninemia in the two groups, whereas changes in neopterin elimination were milder.

A longitudinal study (fig.2), in 2 PKU patients referred to begin their phenylalanine restricted diet, confirmed the parallel changes of pteridine excretion and blood phenylalanine levels.

In blood, similar results were obtained (Table I), the increase of neopterin concentrations in PKU patients was more significant than in urine.

In control subjects following an oral phenylalanine load, the net-changes of pteridine concentrations in urine and serum were not age-related. The results (Table II) showed an increase of biopterin levels which were significantly higher following a 200 mg/kg load than after a 100 mg/kg load. In contrast, the net-change of neopterin levels was not significant.

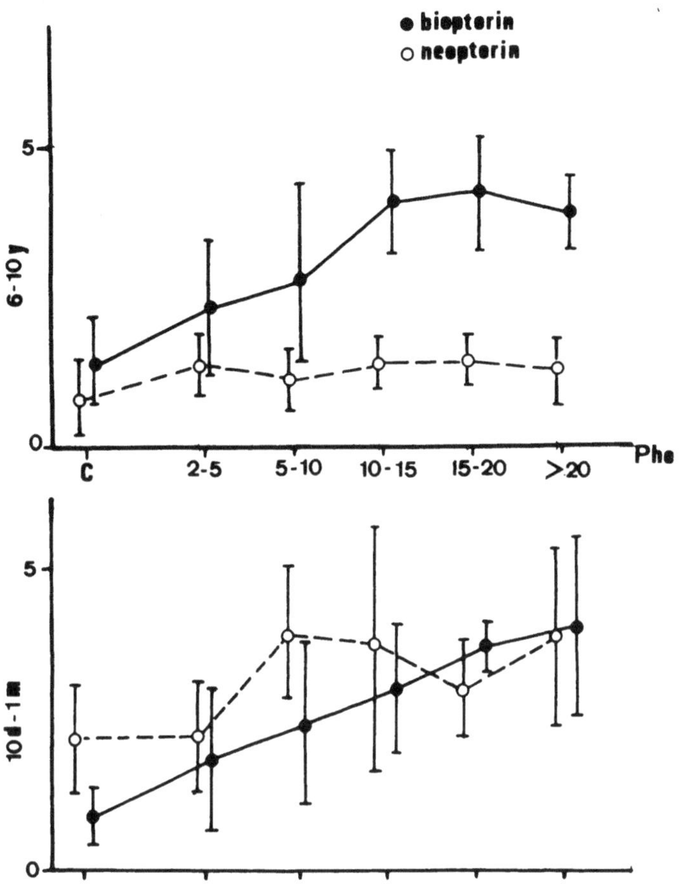

Figure 1

Pteridines excretion (μmol/mmol of creatinine) in controls (C) and PKU patients according to their phenylalaninemia (mg/100ml), in two age-groups (mean ± 1 SD).

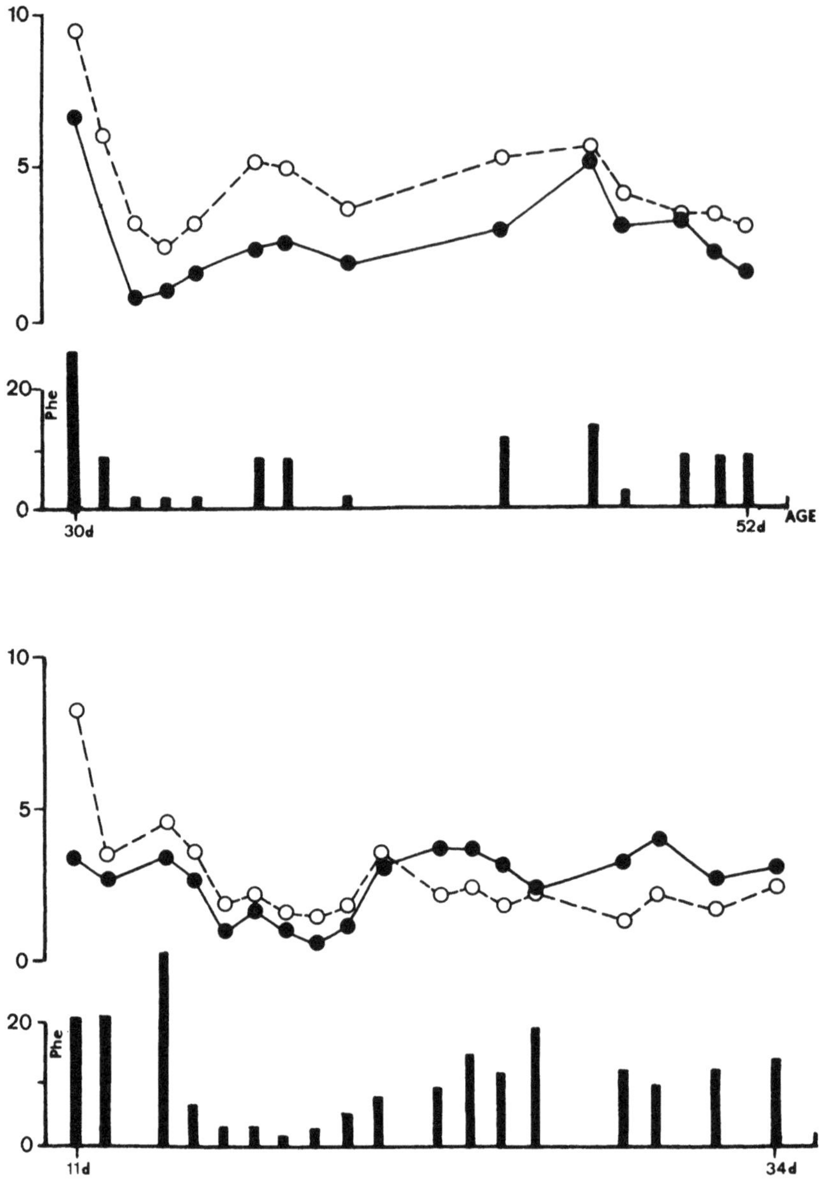

Figure 2

Daily variations of pteridines excretion (µmol/mmol of creatinine) and phenylalaninemia (mg/100 ml) in two PKU babies.
● biopterin
○ neopterin

Table I:

Serum (nmol/l) pteridine concentrations in controls and PKU patients.

		neopterin	biopterin	N/B
15d-2 m	controls (n=14)	44.1 ± 23.2	8.1 ± 3.5	5.7 ± 2.2
	PKU (n=10)	207.9 ± 70.8*	41.5 ± 19.1*	5.6 ± 1.9 ns
4-15 y	controls (n=9)	14.7 ± 3.5	8.2 ± 3.9	2.0 ± 0.8
	PKU (n=9)	46.3 ± 22.9*	28.0 ± 8.5*	1.9 ± 1.2 ns

mean ± 1 SD
ns: not significant
* $p < 0.005$

Table II:

Net increase of neopterin and biopterin in serum of controls following an oral phenylalanine load.

load			net-increase of neopterin	biopterin
100 mg/kg	Serum	(n=17)	1.14 ± 0.33 ns	2.85 ± 1.27 *
	Urine	(n=22)	1.00 ± 0.25 ns	1.70 ± 0.44 *
200 mg/kg	Serum	(n=11)	1.19 ± 0.36 ns	4.31 ± 1.42 *,**
	Urine	(n=13)	1.19 ± 0.31 ns	2.52 ± 0.82 *,**

mean ± 1 SD
ns: not significant
* $p < 0.005$
** $p < 0.005$ when 200 mg/kg loads compared to 100 mg/kg loads.

Table III:

Effect of phenylalanine on serum and liver biopterin in animal models:

A. Rat model:

	BIOPTERIN serum (nmol/l)	BIOPTERIN liver (nmol/g prot)	PHE serum (mmol/l)
Controls	61.4 ± 6.5	61.0 ± 8.1	0.02 ± 0.01
30 min after Phe load (300 mg/kg)	80.9 ± 14.6 *	34.2 ± 8.9 **	1.05 ± 0.33

mean ± 1 SD of 5 experiments.
* $p < 0.05$
** $p < 0.005$

B. Isolated perfused canine liver:

Phe load	1.6 g SERUM Phe *	1.6 g SERUM Biopterin **	1.6 g LIVER Biopterin ***	6.0g SERUM Phe *	6.0g SERUM Biopterin **	6.0g LIVER Biopterin ***
before	0.09	37.0	43.1	0.16	28.9	65.9
20 min after	2.87	50.8	30.5	4.59	111.8	16.3

* mmol/l
** nmol/l
*** nmol/g protein

In the two animal models, the phenylalanine load led to an increase of biopterin concentration in serum, but the liver biopterin content significantly decreased (Table III A and B).

Comments :

The increase of biopterin in serum and urine of PKU patients and controls following a phenylalanine load has been reported by many authors (2-4); but the variations of neopterin concentration were barely investigated.

The present results confirmed the close relationship between phenylalaninemia and pteridine concentrations in biological fluids. However, it appeared that only biopterin raised following a short hyperphenylalaninemic insult, such as in controls following an oral load, whereas neopterin and biopterin increased in PKU patients, a group which represented a chronic hyperphenylalaninemic situation.

The results obtained from the animal models suggested that phenylalanine led to a release of the liver biopterin into the blood (Fig.3b).

In the chronic hyperphenylalaninemic situation, it could be expected a constant out-flow of cellular THB and thus a reduction of the intracellular pool of THB. Recently, some evidence was obtained in our laboratory that a negative feed back control by reduced pteridines (dihydrobiopterin and THB) occured on the biosynthesis of THB at the GTP-cyclohydrolase step (unpublished results). Therefore, a depressed pool of THB might act as the signal for the enhancement of THB synthesis from GTP. However, as the "biopterin- synthetase" step appeared rate-limiting in man, dihydroneopterin- triphosphate could accumulate and, after

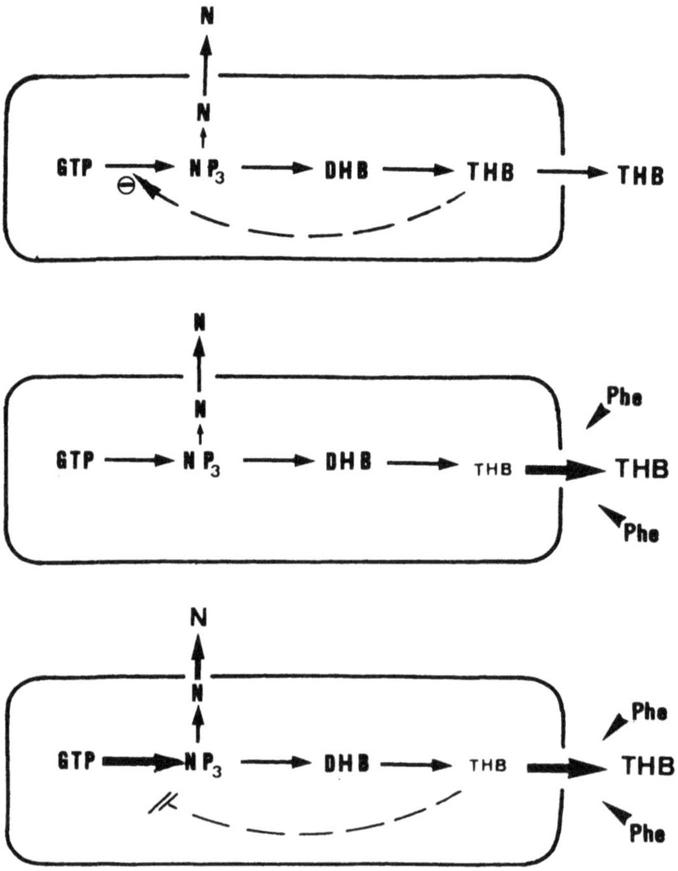

Figure 3 :

A. Normal pathway of the synthesis of Tetrahydrobiopterin (THB). $N\text{-}P_3$: dihydroneopterin-triphosphate, N : neopterin, DHB: dihydrobiopterin, ---▸ negative feed back control of THB on GTP-cyclohydrolase.

B. Effect of a brief phenylalanine load.

C. Effect of a chronic hyperphenylalaninemic status.

dephosphorylation, dihydroneopterin could leave the cell. In regard to such a mechanism, the increase of serum neopterin and biopterin in PKU was explained (Fig.3c). Purdy (7) reported that phenylpyruvic acid was also a potent inhibitor of Dihydropteridine-reductase. Thus in PKU, a second cause of a decreased salvage of THB may occur.

If these mechanisms also concerned the brain, a reduced availabity of THB could limit the monoaminergic neurotransmitter synthesis and partly explained the impaired metabolism of tryptophan often reported in PKU (8,9).

References :

1. Danks,D,M., Bartholome,K., Clayton,B.E., Curtius,H., Grobe,H., Kaufman,S., Leeming,R., Pfleiderer,W., Rembold,H. and Rey,F. Malignant hyperphenylalaninemia - Current status (June 1977). J.Inher.Metab.Dis. 1, 49 (1978)

2. Niederwieser,A., Curtius,H-Ch., Gitzelmann,R., Otten,A., Baerlocher,K., Blehova,B., Berlow,S., Gröbe,H., Rey,F., Schaub,J., Scheibenreiter,S., Schmidt,H. and Viscontini,M. Excretion of pterins in phenylketonuria and phenylketonuria variants. Helv. Paed. Acta 35, 335-342 (1980)

3. Leeming,R.J., Blair,J.A., Green,A. and Raine,D.N. Biopterin derivatives in normal and phenylketonuria patients after oral doses of L-phenylalanine, L-tyrosine and L-tryptophan. Arch.Dis.Child. 51, 771 (1976)

4. Dhondt,J.L., Ardouin,P., Hayte,J.M. and Farriaux,J.P. Developmental aspects of pteridine metabolism and relationships with phenylalanine metabolism. Clin. Chim. Acta 110, 143 (1981)

5. Kestens,P.J. and McDermott,W.V. Perfusion and replacement of the canine liver. Surgery 50, 196 (1961)

6. Dhondt,J.L., Largilliere,C., Ardouin,P., Farriaux,J.P. and Dautrevaux,M. Diagnosis of variants of hyperphenylalaninemia by determination of pterins in urine. Clin. Chim. Acta 110, 58 (1981)

7. Purdy,S.E and Blair,J.A. Rat liver dihydropteridine reductase inhibition. Biochem.Soc.Trans. 8, 565 (1980)

8. McKean,C.M., Schanberg,S.M. and Giarman,N.J. A mechanism of the indole defect in experimental phenylketonuria. Science 137, 604 (1962)

8. McKean,C.M., Schanberg,S.M. and Giarman,N.J. A mechanism of the indole defect in experimental phenylketonuria. Science 137, 604 (1962)

9. Curtius,H.Ch., Baerlocher,K. and Völlmin,J.A. Pathogenesis of phenylketonuria : inhibition of DOPA and catecholamine synthesis in patients with phenylketonuria. Clin.Chim.Acta 42, 235 (1972)

SERUM PTERIDINES IN PATIENTS ON CHRONIC HAEMODIALYSIS

J.L.DHONDT

Laboratoire de Biochimie, Faculté libre de Médecine, 56 rue du port, 59046 LILLE Cédex (France)

Ph.VANHILLE

Service de Néphrologie et Hémodialyse, CHR , 59322 VALENCIENNES Cédex (France)

Previous studies reported alterations in catecholamine metabolism in chronic uraemic patients (1-5). The implication of amino-acids imbalance and/or of endogenous inhibitors have been discussed, but the precise mechanism of these disturbances was not yet completely understood.

One of the rate-limiting steps of the monoaminergic neurotransmitters synthesis is the hydroxylating step: Tyrosine- and Tryptophan-hydroxylases. The cofactor of these enzymes is tetrahydro-biopterin, a pteridine compound synthesized _in vivo_ from Guanosine triphosphate (6). Baker (7) and Leeming (8) first reported the occurence of increased levels of biopterin in blood from uraemic patients. The recent introduction of HPLC methods for the measurement of pteridines in physiological fluids led us to reinvestigate serum levels of unconjugated pteridines in chronic uraemic patients.

Material and Method:

Blood collections: Thirty healthy controls (age 20-40 years) receiving no medication were selected as normal subjects. Sera were taken from 57 patients (20 - 81 years) on maintenance dialysis and from 5 patients with a residual glomerular filtration rate (20-40 ml/mn).

Biochemical and Clinical Aspects of Pteridines · Volume 1
Cancer · Immunology · Metabolic Diseases

Pre- and per-dialysis samples were obtained from six patients.

Pteridine measurements: Sera were immediately separated chromatography of pteridines were performed according to the method previously described (9). Briefly, the pteridines were oxidized in their more fluorescent derivatives by iodine in acidic condition for 1 hour. Excess iodine was then reduced by ascorbic acid and internal standard (6-methyl-pteridine) was added before deproteinization by trichloroacetic acid. The sample was purified through ion exchange chromatographies and the final eluate was directly analyzed by HPLC chromatography, using a cation exchange column (Partisil 10-SCX, Whatman) and a fluorimetric detector.

Results:

The blood levels from healthy adults were 15.9 ± 4.7 nmol/l for neopterin and 6.7 ± 2.4 for biopterin, with a neopterin to biopterin (N/B) ratio of 2.62 ± 1.01. In chronic uraemic patients, pteridine concentrations varied to a large extent from one to the next (Fig.1). In all cases, neopterin values were higher than in controls. Biopterin levels were also significantly increased but not at the same extent, thus raised N/B ratio were observed.

In the patients with moderate chronic uraemia, biopterin was in the normal range, whereas neopterin was increased.

By haemodialysis , close net changes of the pteridine values were observed, but in any case neopterin values were normalized (Fig.2).

Comments:

The finding of high biopterin concentrations in chronic uraemic patients agreed with data obtained by Leeming (8) using the *Chritidia fasciculata* assay. The present results pointed out that a higher increase of neopterin concentrations also occured (up to 75 times the

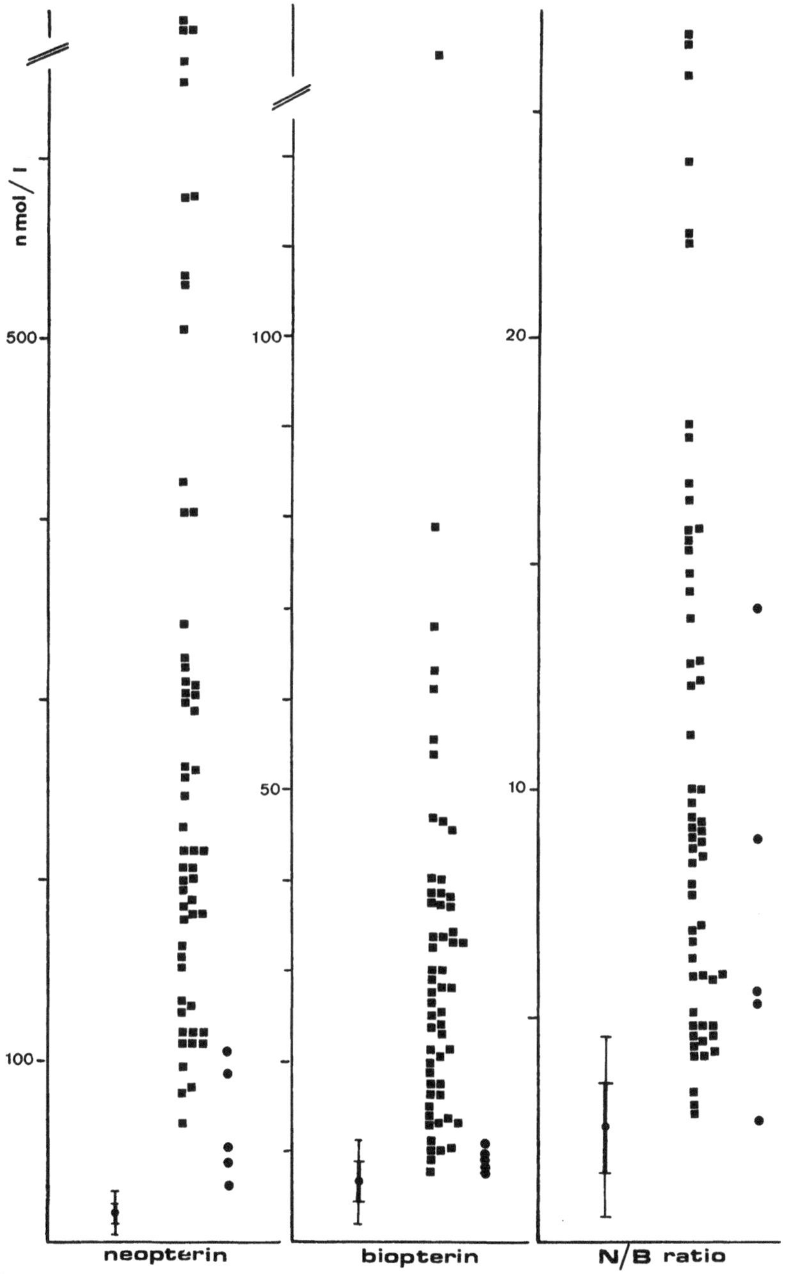

Figure 1

Serum neopterin and biopterin (nmol/l) and molar neopterin to biopterin (N/B) ratio in chronic uraemic patients.
control values (mean ± 1 and 2 SD),
■ patients on haemodialysis,
● patients with residual glomerular filtration.

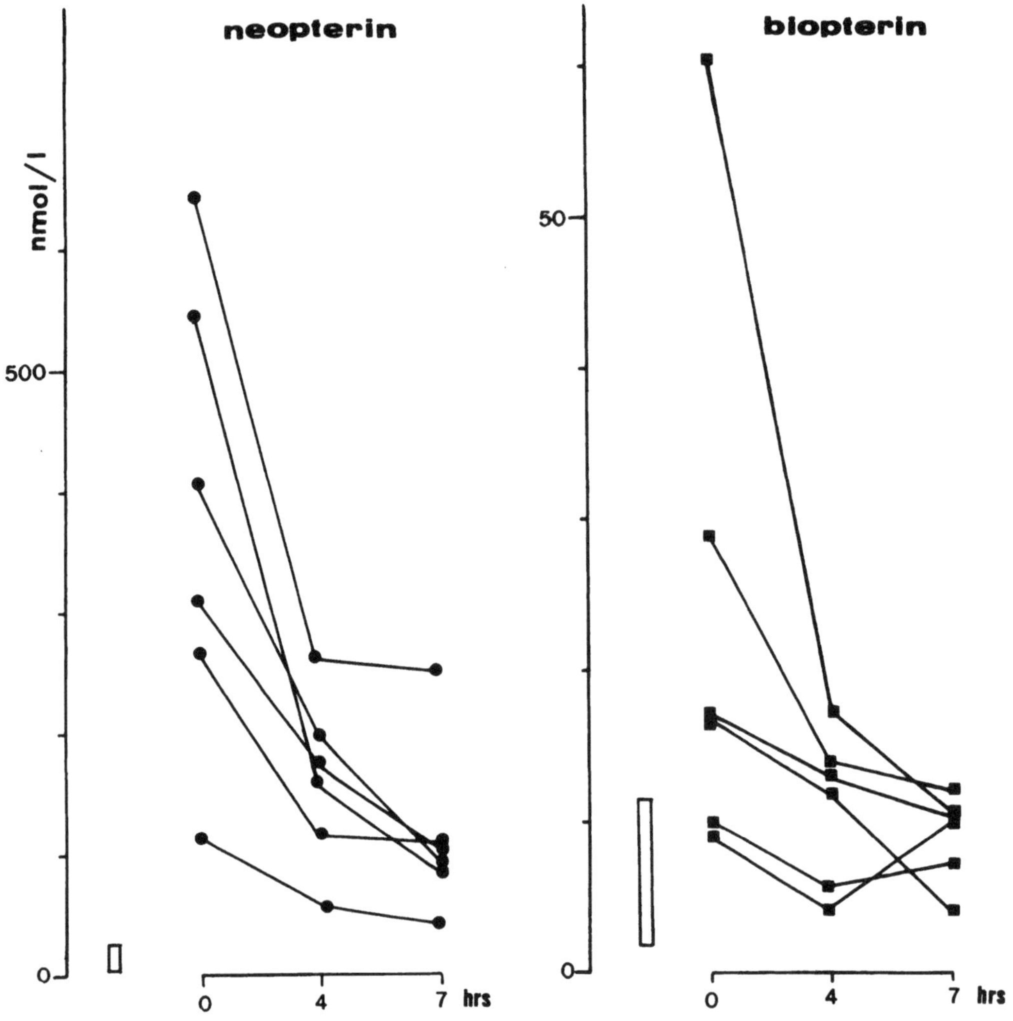

Figure 2

Vraiation of serum pteridine concentrations (nmol/l) in chronic uraemic patients during hemodialysis.
▭ range of normal values (mean ± 2 SD).

normal mean value). As tetrahydrobiopterin (THB) is the cofactor of Tyrosine- and Tryptophan-hydroxylases, these abnormalities took place in the course of the numerous data confirming the impaired metabolism of monoaminergic neurotransmitters in uraemia (1-5).

The increase of serum biopterin levels in chronic uraemic patients could be explained by the renal failure, since the demonstration that kidney played an important role in maintaining biopterin serum concentrations within narrow limits (8,10). In contrast, the very high increase of neopterin levels and high N/B ratio rather reflected an impairement of the THB synthesis.

THB is synthesized in vivo from GTP. Two enzymes appeared to be rate limiting : the GTP-cyclohydrolase and, in human, the Biopterin-synthetasé complexe. The former enzyme seemed to be under a negative feed-back control by reduced pteridines (J.L.Dhondt and Z.Bellhasene, unpublished results) and under hormonal control (11). The latter enzyme complexe exhibit a progressive maturation during childhood (12) and even in the normal adult condition all the dihydroneopterin-triphosphate issued from GTP could not enter this way. As a result, a dephosphorylation occured into dihydroneopterin which could leave the cell and appeared in serum. Finally the formed THB is continuously regenerated through the Dihydropteridine-reductase (DHPR) step.

The present findings of high levels of neopterin and biopterin with a raised N/B ratio in serum chronic uraemic patients were consistent with an increased synthesis of THB associated with a pteridine out-flow, in a similar way that that suggested for catecholamines (4).

The precise mechanism of these features was not yet known. However, the demonstration that DHPR was sensitive to many endogenous and exogenous influences (13-15) raised the possibility of a decrease salvage of THB. A reduced DHPR activity decreases the regeneration of THB (13), which is then rearranged to dihydrobiopterin and appears in serum. Consequently, a reduced pool of THB should be the signal for stimulating the cofactor synthesis at the GTP-cyclohydrolase step; but, since the rate of biopterin synthesis is slow, dihydroneopterin-triphosphate accumulates, is dephosphorylated and also leaves the cell.

A reduced availability of THB limits the monoaminergic neurotransmitter synthesis which could lead to neurological symptoms, as illustrated by observations of inherited THB deficiencies (18).

Even though multiple biochemical events could be responsible of neurologic disturbances in chronic uraemic patients, the present mechanism ought to be further investigated in regard to the therapeutic opportunity offered by the use of THB supplementation (17) and in order to know if individual susceptibility to environmental agents could be related to genetic influences (THB synthesis or DHPR activity).

References:

1. Demassieux,S. Corneille,L., Lachance,S. and Carriere,S. Determination of free and conjugated catecholamines and L-3,4-dihydroxyphenylalanine in plasma and urine : evidence for a catechol-O-methyltransferase inhibitor in uraemia. Clin.Chim.Acta 115, 377-390 (1981).

2. Deferrari,G., Garibotto,G., Robaudo,C., Ghiggeri,G.M. and Tizianello,A. Brain metabolism of aminoacids and ammonia in patients with chronic renal insufficiency. Kidney Int. 20, 505-510 (1981).

3. Corder,C.N., Sharma,J. and McDonald,R.H. Variable levels of plasma catecholamines and dopamine beta-hydroxylase in hemodialysis patients. Nephron 25, 267-272 (1980).

4. Maillet,C.M. and Garber,A.J. The effect of hemodialysis on levels of cyclic nucleotide-associated hormones in patients with chronic renal failure. J.Dial. 4, 121-128 (1980).

5. Sullivan,P.A., Murnaghan,D., Callaghan,N., Kantamaneni,B.D. and Curzon,G. Cerebral transmitter precursors and metabolites in advanced renal disease. J.Neurol.Neurosurg.Psychiat. 41, 581-588 (1978).

6. Häusermann,M., Ghisla,S., Niederwieser,A. and Curtius, H.Ch. New aspects in biopterin biosynthesis in man. FEBS Letters 131, 275-278 (1981).

7. Baker,H., Frank,O., Bacchi,C.J. and Hutner,S.H. Biopterin content of human and rat fluids and tissues determined protozoologically. Amer.J.Clin.Nutr., 27, 1247-1253 (1974).

8. Leeming,R.J., Blair,J.A., Melikian,V. and O'Gorman,D.J. Biopterin derivatives in human body fluids and tissues. J.Clin.Path. 29, 444-451 (1976).

9. Dhondt,J.L., Largilliere,C., Ardouin,P., Farriaux,J.P. et Dautrevaux,M. Diagnosis of variants of hyperphenylalaninemia by determination of pterins in urine. Clin. Chim. Acta 110, 58-60 (1981).

10. Häberle,D.A., Schiffl,H., Mayer,G., Hennings,G. et Rembold,H. Renal balance of pterin cofactors in the rat. Pflügers Arch. 375, 9-16 (1978).

11. Viveros,O.H., Lee,C-L., Abou-Donia,M.M., Nixon,J.C. and Nichol,C.A. Biopterin cofactor biosynthesis : independent regulation of GTP-cyclohydrolase in adrenal medulla and cortex. Science 213,349-350 (1981).

12. Dhondt,J.L., Ardouin,P., Hayte,J.M. et Farriaux,J.P. Developmental aspects of pteridine metabolism and relationships with phenylalanine metabolism. Clin. Chim. Acta 110, 143-152 (1981).

13. Purdy,S.E., Blair,J.A., Leeming,R.J. and Hilburn,M.E. Effect of lead on tetrahydrobiopterin synthesis and salvage : A cause of neurological dysfunction. Intern.J.Envirionmental Studies 17, 141-145 (1981).

14. Leeming,R.J. and Blair,J.A. Dialysis dementia, aluminium, and tetrahydrobiopterin metabolism. Lancet 1, 556 (1979).

15. Purdy,S.E., Blair,J.A. and Barford,P.A. Inhibition of dihydropteridine reductase by dopamine. Biochem.J. 195, 769-771 (1981).

16. Butler,I.J., Koslow,S.H., Krumholz,A., Holtzman,N.A. and Kaufman,S. A disorder of biogenic amines in dihydropteridine reductase deficiency. Ann.Neurol. 3, 224-230 (1978).

17. Kapatos,G. and Kaufman,S. Peripherally administrated reduced pterins do enter the brain. Science 212, 955-956 (1981).

SUBJECT INDEX

AUTHOR INDEX

Bei Fragen zur Produktsicherheit wenden Sie sich bitte an:
If you have any questions regarding product safety,
please contact:

Walter de Gruyter GmbH
Genthiner Straße 13
10785 Berlin
productsafety@degruyterbrill.com